SUBCOMMITTEE ON AVIATION

JOHN L. MICA, Florida, *Chairman*

THOMAS E. PETRI, Wisconsin
JOHN J. DUNCAN, JR., Tennessee
JACK QUINN, New York
VERNON J. EHLERS, Michigan
SPENCER BACHUS, Alabama
SUE W. KELLY, New York
RICHARD H. BAKER, Louisiana
FRANK A. LOBIONDO, New Jersey
JERRY MORAN, Kansas
JOHNNY ISAKSON, Georgia
ROBIN HAYES, North Carolina
TIMOTHY V. JOHNSON, Illinois
DENNIS R. REHBERG, Montana
SAM GRAVES, Missouri
MARK R. KENNEDY, Minnesota
BUD SHUSTER, Pennsylvania
JOHN BOOZMAN, Arkansas
CHRIS CHOCOLA, Indiana, *Vice Chairman*
BOB BEAUPREZ, Colorado
STEVAN PEARCE, New Mexico
JIM GERLACH, Pennsylvania
MARIO DIAZ-BALART, Florida
JON C. PORTER, Nevada
VACANCY
DON YOUNG, Alaska
(Ex Officio)

PETER A. DEFAZIO, Oregon
LEONARD L. BOSWELL, Iowa
WILLIAM O. LIPINSKI, Illinois
JERRY F. COSTELLO, Illinois
ELEANOR HOLMES NORTON, District of Columbia
ROBERT MENENDEZ, New Jersey
CORRINE BROWN, Florida
EDDIE BERNICE JOHNSON, Texas
JUANITA MILLENDER-MCDONALD, California
ELLEN O. TAUSCHER, California
BILL PASCRELL, JR., New Jersey
TIM HOLDEN, Pennsylvania
SHELLEY BERKLEY, Nevada
BRAD CARSON, Oklahoma
JIM MATHESON, Utah
MICHAEL M. HONDA, California
RICK LARSEN, Washington
MICHAEL E. CAPUANO, Massachusetts
ANTHONY D. WEINER, New York
NICK J. RAHALL II, West Virginia
BOB FILNER, California
JAMES L. OBERSTAR, Minnesota
(Ex Officio)

USE OF BIOMETRICS TO IMPROVE AVIATION SECURITY

(108–69)

HEARING

BEFORE THE

SUBCOMMITTEE ON
AVIATION

OF THE

COMMITTEE ON
TRANSPORTATION AND
INFRASTRUCTURE
HOUSE OF REPRESENTATIVES

ONE HUNDRED EIGHTH CONGRESS

SECOND SESSION

MAY 19, 2004

Printed for the use of the
Committee on Transportation and Infrastructure

U.S. GOVERNMENT PRINTING OFFICE

95–133 PDF WASHINGTON : 2005

For sale by the Superintendent of Documents, U.S. Government Printing Office
Internet: bookstore.gpo.gov Phone: toll free (866) 512–1800; DC area (202) 512–1800
Fax: (202) 512–2250 Mail: Stop SSOP, Washington, DC 20402–0001

COMMITTEE ON TRANSPORTATION AND INFRASTRUCTURE

CONTENTS

TESTIMONY

PREPARED STATEMENT SUBMITTED BY A MEMBER OF CONGRESS

PREPARED STATEMENTS SUBMITTED BY WITNESSES

ADDITION TO THE RECORD

USE OF BIOMETRICS TO IMPROVE AVIATION SECURITY

Wednesday, May 19, 2004

HOUSE OF REPRESENTATIVES, SUBCOMMITTEE ON AVIATION, COMMITTEE ON TRANSPORTATION AND INFRASTRUCTURE, WASHINGTON, D.C.

The subcommittee met, pursuant to call, at 10:05 a.m. in room 2167, Rayburn House Office Building, Hon. John L. Mica [chairman of the subcommittee] presiding.

Mr. MICA. Good morning. I would like to call the meeting of the Aviation Subcommittee to order.

This morning's hearing is going to focus on the use of biometrics to improve aviation security, and also review some of the progress of instituting biometric standards with the Department of Homeland Security and other Federal agencies.

We have two panels. I am going to ask all of the panelists to come up today and join us. We will hear from the Government panel first, and the second panel right afterwards. Somehow they want to be separated. But we will get them together here. I am a uniter, not a divider.

We are going to start with opening statements from Members. So we will start this morning's hearing with my opening statement, and then I will yield to other Members as they join us.

Again, the purpose of today's hearing is to review the progress, or lack of progress, in using biometric technologies to improve aviation security, and also to examine how similar biometric standards are being incorporated into our existing security systems.

We are spending billions of dollars each year to screen passengers and bags for weapons and explosives. But after some two and a half years since September 11, we have failed to adopt a biometric standard to address the even more basic problem of airport access control. The low tech security credentials that are currently being used to authorize access to the most sensitive areas of our Nation's aviation system could be courting disaster. Our multibillion screening regime is defenseless against a terrorist who uses a lost, stolen, or forged security badge or law enforcement officer credential to walk right past a screening checkpoint.

This Committee took action on this issue in the Aviation Transportation Security Act, which was enacted shortly after the terrorists acts of September 11. I have got up on the screen the provisions of this Act dealing with biometrics. The Act includes several provisions intended to strengthen airport access control through the use of biometrics. I look forward to hearing from our witnesses today regarding the actions that have been taken in response to these provisions, or the lack of response. And again, we have various sections of the bill that refer to these requirements or directives, that language put into the Act, we want to hear about progress on.

Unfortunately, we have a hodgepodge of airport security credentials today and access control systems at our Nation's airports. Each individual airport is responsible for its own security and issues its own security badges. These badges are referred to as "sterile" and given the acronym SIDA badges. To date, they contain no biometric standard. The sterile badges authorize access to the terminal areas beyond screening checkpoints.

Biometrics could improve employee, passenger, and flight crew identify verification and access authorization. For example, adding biometrics to existing access control systems could protect against unauthorized accessing using lost, stolen, or forged badges. Biometrics could also protect against a terrorist on a Watch List attempting to obtain a credential using an assumed identity. In addition, biometrics could protect against the impersonation of a pilot, other crew member, air traffic controller, or employee of the airport.

Biometrics could also help close a similar gaping hole in our aviation security system having to do with law enforcement credentials. Currently, law enforcement officers armed with a weapon can fly at any time simply by presenting their agency's credential. In fact, law enforcement officers from 18,000 separate State and local law enforcement agencies may fly armed if they present their agency's credential and a letter on their agency's letterhead stating they have an official work related reason to fly armed. And this does not include the multitude of Federal law enforcement personnel.

Lost, stolen, or forged law enforcement credentials could easily be used by unauthorized persons to carry guns on board. I just asked one of my sheriffs to show me what they need to present. So they write a letter, and this is a copy of the credentials that the Volusia County Sheriff's Office is required to present. Now let me say, I have no problem with officers carrying weapons on board from some 18,000 agencies or Federal officers. The problem I have is having some biometric standard to ensure that that is the individual who is authorized. I want to make that perfectly clear.

Fake badges of all kinds are widely available on the Internet. In just ten minutes worth of research, one of my Subcommittee staffers located the following sources on the web. They have got it up on the screen here. Slide one, badge stuff.com is where you can buy all the equipment and software you need to make your own badges. Slide two, at fire store on line.com you can buy a sheriff's badge for $39.99. Slide three, make your own fake IDs.com is where you can have an ID card tailored to your specifications. Slide four, select from 1,816 different ID card logos, including many police department logos. Slide five, buyidentity.com boasts of the most advanced technology, including features such as bar codes, smart chips, and overlay holograms. So, I was a little bit shocked by what we found. And this is not to degrade the talents of any college student who can make great IDs. We were going to display some of those but the local bars wanted to hold on to them.

Using biometrics may be the only way to ensure that the person presenting the law enforcement or other credentials is actually entitled to that credential. Even more startling is that GAO conducted an undercover test on exactly this issue in the year 2000. GAO agents created fake law enforcement identification using commercially available software packages and information downloaded from the Internet. The agents then used the faked credentials to perform penetration tests at various Federal buildings as well as two commercial airports. The GAO agents were 100 percent successful in penetrating each site.

At the two airports GAO visited, the agents used tickets that had been issued in their undercover names. These agents declared themselves to be armed law enforcement officers, displayed their counterfeit badges and identification, and were issued law enforcement boarding passes. The GAO agents then presented themselves at the security checkpoints and were waived around the magnetometers. Neither the agents nor their briefcases were screened. A copy of the GAO report on this penetration test was later found in an Al Qaeda cave in Afghanistan we have learned. I have got a copy of testimony, this is not the report but is testimony, before the Subcommittee on Crime of the House Committee on Judiciary, dated May 2005, and I will submit that for the record. Without objection, so ordered.

Let me also say, I recognize that biometrics may not be a total panacea. Biometrics can tie an identity to a particular person, but biometrics alone cannot ensure that the identity is always accurate or that the person is not a potential terrorist. Furthermore, biometric systems are not 100 percent accurate, and certain systems may be vulnerable to intentional thwarting. For example, some fingerprint systems, I am told, can be thwarted by the use of a fingerprint impression on a gummy bear. Not very sophisticated, but I understand it works.

Two and a half years later and approaching the three year anniversary of September 11, we still lack performance data on many biometrics. Without large data samples to use for testing and methodology standards for such testing, it is difficult to evaluate each of the many vendors' claims of accuracy. Even with these caveats, however, I believe the use of biometrics still has the potential to significantly improve aviation security. Again, it has been some two and a half years, going on three years since September 11th and we know that airport access remains of interest to terrorists. According to the Transportation Security Administration, within the past few months alone there have been several suspicious incidents of possible surveillance of airports, including surveillance of an area containing a SIDA access door.

We need to address this issue without delay. I am kind of surprised that nearly three years later we still do not have a biometric standard adopted at the Federal level that can be used as a model for Federal identification for State and local and other agencies with some certainty of properly identifying the individuals who carry those credentials.

I look forward to hearing the testimony of our witnesses on this important topic and discussing the ways in which biometrics may be used to close the gap in our security system.

I would like to yield at this time to the Ranking Member, Mr. DeFazio.

Mr. DEFAZIO. Thank you, Mr. Chairman, and thank you for your persistence in this matter. Unfortunately, it seems that the Transportation Security Administration is continuing to move at a glacial pace on these issues. For the life of me, I cannot understand why we have not been able to develop and issue a uniform national transportation worker identification card. It is beyond me. How could this take years? We have this mishmash of badges, some of which are flashed at rather inattentive security guards, as we noticed at Detroit airport, to gain access to the terminal without any minimal screening while you are wearing a bulky coat and carrying big bags full of whatever. This is security? This is just extraordinary to me.

We have created the illusion of security for the American people. It manages to inconvenience them, put them in long lines, but they all say, everyone I talk to, "I do not mind, it is making us safe." But if they knew people were going around the system—and I almost had the briefing two weeks ago to answer the question I have been asking for more than a year, how many airports allow vendors and other people who work in the airport to freely go in and out daily without going through any security at all? But for some reason, the briefing was canceled, so I have not had it. So we do not have uniform identification cards. In some airports, how many hundreds of thousands a people a day, we do not know, people flash these non-uniform cards at inattentive guards and walk through, while the pilots, the flight attendants, and all the passengers are over there standing in long lines to go through security.

These are not unsophisticated people we are dealing with in any fashion. They tested the system again and again and again before they struck on September 11. The Chairman displayed the GAO report. He showed you that people can go on line and buy fake IDs. Why can we not get to a uniform national ID for transportation workers, for anybody who is getting access to secure areas in airports? I think it is a reasonable question. And why has it taken more than two years? I think the technology exists, I think there are plenty of models out there in the private sector, in other Government sectors. And then biometrics, not only doing the background checks on the people to be certain they are who they say they are, but then issuing them a card which will verify that they are the person to whom that card was issued. Again, the technology has existed for years. It is being used elsewhere in the private sector. It is being used by other levels of government in this country, other governments around the world. Why can't the United States of America, the country who was attacked, put in place such a system? Why are we taking this sort of lackadaisical attitude toward this? I just do not understand.

And finally, I am pleased to see, after the Chairman and I for two years have been raising the issue of trying to reduce the burden on the screeners, expedite passengers through the airports, and help the airlines with some of their highest revenue customers, we are finally, after two years, moving ahead with what has been called a number of things, but a trusted or Registered Traveler Program, where we will be able to expedite people whose backgrounds have been vetted, who will be issued, hopefully, I am not sure what you can issue them, some sort of a biometric identification, through security lines. This could help the airlines, it could help the screeners and security by allowing them to focus on unknown people, and it will be an improvement in the system. I am pleased the pilot is moving forward. I will be looking forward to understanding and being briefed on that.

But on these other issues, I have got to say, time and time again the Chairman has convened meetings—privately, publicly, secure and unsecure—and we have raised these issues time and time again, and here we are two years later and we are still talking about it. I fear that someday this is going to have catastrophic consequences. We have got to get this done. Thank you, Mr. Chairman.

Mr. MICA. I thank the gentleman. Other opening statements? Mr. Matheson?

Mr. MATHESON. Thanks, Mr. Chairman. Just real briefly. It is real clear that we have got to do better at securing our airports. We have got to embrace technology and we have got to embrace biometrics.

I introduced the Aviation Security Technology Enhancement Act on October 11, 2001, along with my colleague Congressman Honda. That legislation called for the establishment of best practices for emerging aviation security technologies, and it created a pilot program for the FAA to test new and emerging aviation technologies in at least 20 airports across the country. Our legislation would have taken the necessary steps to examine the effectiveness and cost of security technologies, including biometrics, in our Nation's airports.

I was pleased that many of the provisions of that legislation were included in the aviation security legislation that was passed by Congress and signed into law almost three years ago. The provision was in the item that was up on the screen at the start of this hearing a few minutes ago. Specially, there is a provision that established demonstration projects at 20 airports nationwide to evaluate emerging security technology. So I look forward to hearing from today's witnesses about the status of these demonstration projects, which I remain hopeful that efforts from those projects will pave the way for more efficient and effective security in airports throughout the country. I yield back.

Mr. MICA. I thank the gentleman. Any further opening statements? If not, we will go right to our panel of witnesses. We will recognize first the Honorable Stewart Verdery, Assistant Secretary for Policy, Border and Transportation Security, U.S. Department of Homeland Security. We also have another Government witness, which is Keith Rhodes, Chief Technologist, Applied Research and Methods, U.S. General Accounting Office. So we will hear from the two Government panelists first.

Stewart Verdery, Assistant Secretary for Policy, Border and Transportation Security of DHS. Welcome, sir, and you are recognized.

TESTIMONY OF HON. STEWART VERDERY, ASSISTANT SECRETARY FOR POLICY, BORDER AND TRANSPORTATION SECURITY, U.S. DEPARTMENT OF HOMELAND SECURITY; AND KEITH A. RHODES, CHIEF TECHNOLOGIST, APPLIED RESEARCH AND METHODS, U.S. GENERAL ACCOUNTING OFFICE

Mr. VERDERY. Thank you, Chairman Mica, Mr. DeFazio, and other distinguished members of the Committee, it is a pleasure to appear before you today to discuss how the Department of Homeland Security is using biometrics to enhance aviation security and also to facilitate legitimate trade and travel. This is my first opportunity to appear before this Subcommittee and I hope that in my role as Assistant Secretary as Border and Transportation Security Policy and Planning, I will be able to explain how we are using biometrics to enhance the security and facilitation missions assigned to the BTS directorate and our agencies.

Biometrics is the science of identifying, recording, and matching unique physical characteristics, such as fingerprint, facial, iris, or hand, to particular individuals. The ability to verify and freeze an individual's identity in this manner has numerous applications for improving the security and efficiency of our transportation and immigration systems.

The Department, under the leadership of our Science and Technology Directorate, is assessing future applications for biometrics and also examining how to leverage the success of existing programs. One of the principal reasons for having a BTS Directorate is to oversee programs run by our bureaus, that is Transportation Security Administration, Customs and Border Protection, and US-VISIT, among others, is to find synergies and to apply lessons learned across multiple program offices.

One of these offices, of course, is US-VISIT, which has been deployed successfully while meeting its goals of enhancing security, facilitating travel and trade, ensuring integrity of our immigration systems, and protecting privacy. US-VISIT is adding an average of only 15 seconds to the inspection process, yet is significantly enhancing the security of travellers by collecting biometric and biographic information, comparing that information with data collected by the Department of State at the time of visa issuance, and vetting the biographic and biometric information against Watch Lists and other criminal history information.

Today, more than 130 of the 211 visa-issuing posts overseas are capturing fingerscans and photographs of foreign nationals when they apply for visas. At the U.S. border, visitors provide this biometric as well as biographic and other documentation which is checked against the US-VISIT databases, including visa issuance information, terrorist watchlists, and immigration status information. In its first four months of operation, DHS has processed over 4 million foreign applicants for admission through US-VISIT at our air and sea ports of entry. During that period, approximately 340 individuals were identified by biometrics alone as being the subject of a lookout, some 60 percent for criminal violations. Among the many hits was a drug dealer who had entered the U.S. more than 60 times in the past four years using multiple names and dates of birth, before being detected on his first trip under US-VISIT.

The Department is also exploring the use of biometric technology to better secure sterile areas in airports. TSA has commenced Phase I of the Airport Access Control pilot program. One project, for example, will test a system that combines fingerprint biometrics and RFID technology to control vehicle access. Another will test a system that uses fingerprint biometric technology to allow only authorized persons to enter secure areas of an airport. And yet a third will control access to the secure area via an iris biometric recognition system.

In a fusion of access control and identification purposes, TSA has been testing alternatives for developing and/or implementing a secure credential for transportation workers through the TWIC program. These credentials could be used to mitigate potential threats posed by workers in the transportation sector with fraudulent identification. The program is intended to enhance security controls applicable to personnel whose duties require unescorted access to secure areas of a transportation system.

An implemented TWIC program would most likely incorporate the use of biometrics to identify each TWIC holder as unique, linking individuals to their cards and to their security assessment. Biometrics could also be incorporated locally as part of a physical and/or logical access control, leveraging the existing local facility control systems to the maximum extent possible.

The current phase of TWIC includes testing at a variety of transportation facilities, a complete cost benefit analysis, and a review of security effectiveness. And biometrics will clearly be a key component of whatever shape the TWIC program takes in the future.

Now concerning the Registered Traveler Pilot Program, as mentioned in the opening statements, I share the Subcommittee's keen interest in establishing an RT program that will attract travelers and allow TSA to better utilize its resources.

TSA's pilot testing for a Registered Traveler program is designed to determine the feasibility of providing expedited movement through airport security checkpoints for travelers who volunteer personal information and receive a clean security assessment. Volunteers who participate in such an RT pilot program will be required to submit personal data, such as biometrics, that will be used for identity verification. Participants in the program will still be required to submit to screening for weapons, explosives, and prohibited items at the checkpoint. And in June, TSA will begin RT pilots at a limited number of airports which will last for approximately 90 days.

Part of the RT pilot program will focus on improving law enforcement officer LEO credentials. The use of so many different types of LEO credentials increases the risks of an unauthorized armed person could use the forged credential to board an airplane. Under the RT pilot, LEOs who wish to fly armed at the five pilot airports will be issued a biometric identification card by TSA to ensure that the individual seeking to carry a gun on board is, in fact, authorized to do so by the LEO's parent agency.

In conclusion, the advent of automated matching capability gives us the ability to improve performance and permit the deployment and use of new biometric technologies to assist us in freezing or fixing the identities of foreign nationals, improving document security, and deterring illegal access. In order to maximize our return on investment, it is vital that Federal agencies and associated industries who are also responsible for security of infrastructure work together to create compatible systems which will bring our Nation's transportation and immigration security systems into the 21st century. Technology must be utilized in achieving our goals of secure U.S. borders and transportation systems and open doors to legitimate trade and travel.

I thank you for the opportunity to be here. I look forward to your questions.

Mr. MICA. Thank you, and we will hold questions.

We will hear next from Keith Rhodes, Chief Technologist, Applied Research and Methods, of the U.S. General Accounting Office. Welcome, sir, and you are recognized.

Mr. RHODES. Thank you. Chairman Mica, Ranking Member DeFazio, and members of the Subcommittee, I appreciate the opportunity to participate in today's hearing on the use of biometrics for aviation security.

Technologies called biometrics can automate the identification of people by one or more of their distinct physical or behavioral characteristics—by something they are. The term biometrics covers a wide range of technologies that can be used to verify identity by measuring and analyzing human characteristics. Biometrics theoretically represent a more effective approach to security because each person's characteristics are thought to be distinct and, when compared with identification cards and passwords, are less easily lost, stolen, or guessed.

When used for personal identification, biometric technologies measure and analyze human physiological and behavioral characteristics. Unlike conventional identification methods that use something you have, such as an identification card to gain access to a building, or something you know, such as a password to log on to a computer system, these characteristics are integral to something you are.

Biometric technologies vary in complexity, capabilities, and performance, but all share several elements. Biometric identification systems are essentially pattern recognition systems. They use acquisition devices such as cameras and scanning devices to capture images, recordings, or measurements of an individual's characteristics and computer hardware and software to extract, encode, store, and compare these characteristics. Because the process is automated, biometric decision-making is generally very fast, in most cases taking only a few seconds in real time.

Depending on the application, biometric systems can be used in one of two modes—verification or identification. Verification, also called authentication, is used to verify a person's identity; that is, to authenticate that individuals are who they say they are. Identification is used to establish a person's identity; that is, to determine who a person is, regardless of who they say they are. Although biometric technologies measure different characteristics in substantially different ways, all biometric systems start with an enrollment stage followed by a matching stage that can either use verification or identification.

Biometrics is a relatively young technology, having only recently reached the point at which basic matching performance can be acceptably deployed. It is necessary to analyze several metrics to determine the strengths and weaknesses of each technology and vendor for a given application. The effectiveness of any biometric system is a balance between (1) the false match rate—that is, how many times someone is incorrectly identified as being someone else; (2) the false non-match rate—how many times someone is not identified as who they are; and (3) failure to enroll rate—how many times people are not able to enroll in the system for whatever reason.

Identifying, exchanging, and integrating information from different and perhaps unfamiliar sources and functions are essential to an effective biometrics application. Without standards, system developers may need to define in detail the precise steps for exchanging information, a potentially complex, time-consuming, and very expensive process. Progress has been made in developing biometrics standards; however, the majority of biometric devices and their software are still proprietary in many respects. For example, the method for extracting features from a biometric sample, such as a fingerprint, differs among most, if not all, vendors. Devices from company A do not necessarily work compatibly with devices from companies B and C.

As you have heard, the FAA, and subsequently DHS and TSA, have been examining the use of biometrics for aviation security for several years. They, with the Department of Defense, examined the use of biometrics in four aviation security applications: (1) identity verification of employees and ensuring that access to secured areas within an airport is restricted to authorized personnel; (2) protection of public areas in and around airports using surveillance; (3) identity verification of passengers boarding aircraft; and (4) identity verification of flight crews prior to and during a flight. In 2002, TSA contracted with the International Biometric Group to evaluate the use of biometrics for automated surveillance within airports, trusted traveler cards for passengers, and identity verification of employees for access control in airports.

And as you have stated, Mr. Chairman, since the 2001 terrorist attacks, the Congress has directed a greater use of biometrics. While biometric technology is currently available and used in a variety of applications, questions do remain regarding the technical and operational effectiveness of biometric technologies in large-scale applications. We have found that a risk management approach can help define the need and use for biometrics for security. Biometric technologies are available today that can be used for aviation security. However, it is important to bear in mind that effective security cannot be achieved by relying on technology alone. Technology and people must work together as part of an overall security process. As we have pointed out, weaknesses in any of these areas diminish the effectiveness of the security process.

We have found that three key considerations need to be addressed before a decision is made to design, develop, and implement biometrics into a security system: (1) Decisions must be made on how the technology will be used; (2) a detailed cost-benefit analysis must be conducted to determine that the benefits gained from a system outweigh the costs; and (3) a trade-off analysis must be conducted between the increased security, which the use of biometrics would provide, and the effect on areas such as privacy and convenience.

Mr. Chairman, this concludes my statement. I would be pleased to answer any questions that you or members of the Subcommittee may have.

Mr. MICA. Thank you. I will withhold questions. I want to hear from the other two panelists, Mr. Richard Norton, Executive Vice President of the National Biometric Security Project; and Mr. Martin Huddart, Chairman of the Board of Directors of the International Biometric Industry Association.

We will first hear from Richard Norton, with the National Biometric Security Project.

TESTIMONY OF RICHARD E. NORTON, EXECUTIVE VICE PRESIDENT, NATIONAL BIOMETRIC SECURITY PROJECT; AND MARTIN HUDDART, CHAIRMAN, BOARD OF DIRECTORS, INTERNATIONAL INDUSTRY ASSOCIATION

Mr. NORTON. Good morning, Mr. Chairman, and members of the Subcommittee. Thank you for the opportunity to testify on behalf of the National Biometric Security Project, or NBSP. NBSP is one clear sign of Congress' commitment to establish comprehensive new security capabilities in the wake of September 11th--to plug the sort of holes in our system that have been noted this morning--by developing sophisticated advanced biometric technologies solutions to address this specific need.

The specific mission of the NBSP is to provide the Federal Government with the R&D capabilities, the testing and evaluation capabilities, and the deployment experience to get these broad based solutions in place and address these gaps in the critical infrastructure. We are, therefore, concentrating in four key areas to address this problem. The first area is to sort of conduct the applied research necessary to know what our requirements are. Second is to establish a cadre of trained professionals who can deploy these solutions, the scientists, the engineers, and the experts, of which there are only a handful today, by creating educational programs to provide that level of expertise. Third, we are establishing a testing and evaluation capability, as noted by both the Department of Homeland Security and GAO, to make sure that vendors' claims are accurate, and to make sure that the equipment will perform as expected in the rigorous operational environments that we often face. Conducting trials in laboratories is one thing. Making sure they operate under a variety of ambient conditions and over a long period of time is another story. And finally, we have been moving forward aggressively in the area of standards to make sure that we are adequately represented and push the United States standards agenda in every forum possible to make sure we have the interoperability needed to make these systems work.

We have a lot of lessons we have already learned about how effective biometrics can be, especially within an airport footprint. A program has run at San Francisco for the better part of a decade now that has proved that you can manage a system, that it is a cost effective system, and that it can help protect your airport with your employees, making sure that only the authorized people have access to sensitive areas such as ramps and the jetways. That program, as I said, has been in place for ten years.

However, biometrics have not expanded much beyond that extent, for a variety of reasons. The Federal aviation regulations noted the need to identify people securely. And as we have seen today, merely having an ID badge in your hand does not provide that level of security. But biometrics were not mandatory and no clear mention was made of the desire to move in that direction until September 11th. Now a number of efforts are underway to address that problem.

TSA's Airport Access Control Program is one of those. It should help supplement the information we already have on sites such as San Francisco with good metrics on how other biometrics can get measurements of usability of these biometrics. We think that between the San Francisco experience and the TSA trials, a clear case will be made that an airport footprint can be protected using current biometric technology. The bigger challenge is going to be to move forward with a national program that can cover the itinerant workers in the system, the people who are not just employees of a particular airport, but move about in the course of the their business, such as airline employees who will show up at a variety of different airports, often multiple airports, in a single day. To address this, we are looking at a number of things to prime the pump: Again, interoperability standards; taking steps to ensure that the problem of multiple identities can be corrected; and ensuring that the systems developed recognize cost factors and certain privacy concerns about sharing of the information.

The Transportation Worker ID Card Program we think provides a solid structure for addressing this problem. TWIC has come up with a good concept of operations where people will be screened for multiple identities using fingerprints and face recognition technologies. However, applying those same technologies might not be the most appropriate way to protect an airport. We think the TWIC architecture has taken into account the need to be able to upgrade the system, adopt new technologies as they are introduced, and apply the best biometric technology to the particular operational need. TSA is also examining the existing infrastructure to try to make sure that we leverage existing capabilities. NBSP thinks that this is an absolute essential compliment of the approach to make sure that we can introduce such a system in a cost effective way.

NBSP is taking a couple of different actions on a variety of different levels to make sure that we have the testing and research support that is available to the Government as they move forward with deployment. For example, Mr. Chairman, you noted the requirement for a database of information that can help with research. We are working actively with TSA and NIST in establishing a database of fingerprint, iris, and face recognition templates that can be used for further research to make sure that these technologies work properly in the field.

Again, we are establishing a laboratory that can provide the sort of testing regime that can evaluate the effectiveness of biometric products and solutions prior to their installation in the field. And we are developing a cadre of trained biometric professionals who not only can do the testing, but also be of assistance in developing the requirements and supporting deployment.

We are also working with the Customs and Border Protection Bureau and also US-VISIT Program Office to identify citizens who are coming into the country in an effective way, kind of a registered traveler program on behalf of U.S. citizens, who are often the forgotten component at the borders and often are now forced to wait in long lines.

So we are working actively with the Administration on a number of fronts to provide these capabilities, as directed by Congress, and would certainly welcome any questions about the efforts of NBSP to date. Thank you.

Mr. MICA. Thank you.

We will hear now from Martin Huddart, who is Chairman of the Board of Directors of the International Biometric Industry Association. Welcome, and you are recognized.

Mr. HUDDART. Thank you, Mr. Chairman, and the Subcommittee for the opportunity to speak on behalf of the International Biometric Industry Association.

Biometrics have had a long history in both the public and private sectors of protecting critical national infrastructure, computer networks, preventing welfare fraud, border control, and even labor management. I think when restaurant workers in McDonald's are punching in for work using biometrics, I am comfortable that this is no longer new and emerging technology.

Since 9–11 there have been many initiatives with biometrics. However, two and a half years later, there has been little done to implement the technologies outside of the US-VISIT program. Fifteen years ago, the Department of Energy of the U.S. Government decided that credentials alone were not sufficient to protect our Nation's nuclear facilities. They decided that a credential had to be tied to an individual using biometrics. So today, and for the past 15 years, hand readers have protected our Nation's nuclear facilities, 97 percent of those facilities in this country.

Looking at the aviation segment, current TSA regulations imply the need for biometrics. For example, regulations state "only those individuals authorized to have unescorted access" should have access. The regulations do not say authorized pieces of plastic, it says "authorized individuals." That implies the need for biometric technology to meet current regulations in aviation. Cards are just pieces of plastic that can be used by other people. For example, a colleague of mine was told confidentially by a Category X airport in the United States—that is one of the top 20 airports—that in a given year 400 badges go missing at that airport and are replaced. I think that represents a significant security issue.

There is a long track record of biometrics in aviation in the United States. Mr. Norton mentioned San Francisco Airport has been doing this since 1991. Today, 15,000 workers at San Francisco Airport will use biometrics to get to the air operations area. In total, to our best knowledge, there are about eleven airports with significant deployments of biometrics today in the United States. Out of a total of 429, that represents less than 3 percent using hand geometry, fingerprint, and iris technologies in airports. Compare that 3 percent in aviation with 97 percent in the nuclear industry.

It is ironic that some of the well-intentioned efforts that came out after 9–11, and I personally spent time with Congressman Matheson and Congressman Honda on some of these initiatives, some of them, unfortunately, have actually I believe slowed progress down. For example, the pilot program to implement biometrics at 20 airports. We are now just entering the first eight airports to be installed with the pilots and I think we are many, many months away from a conclusive report. And it is unclear to me what the outcome of that effort will be. Curiously, some of the more proven technologies have been excluded from this testing so far as we focus on new and emerging technologies, not proven technologies. Through my own calculations, I estimate that the money spent on this pilot effort could have retrofitted by now 45 of the top 200 airports with biometric technology.

Turning now to the Transportation Workers ID Credential. I think this is a very positive program and it provides a clear infrastructure for implementing biometric identity verification within the airport community. I think it makes an intelligent distinction, as all biometrics are not created equal, as has been said earlier. Some biometrics are better suited as a reference biometric that helps the enrollment process. This makes sure that people do not apply for multiple credentials under different identities and it enables background checks. That is an important part of the process. But the TWIC program also references the operational requirements. When you have large volumes of people coming through access points, fast and efficient verification is a different operational requirement for biometrics, and they have done a good job of distinguishing that. We recommend that this program proceeds quickly and is funded to be implemented very quickly.

We also welcome the recent solicitation for a Registered Traveler Pilot Program, another important application. This is just a great application of biometrics, because it not only improves security but provides better convenience on behalf of the traveler. What we encourage, though, is a look at existing implementations. Again, there is a focus on pilot systems and testing. Let us look at what has already been implemented. Since 1993, the INS implemented N-PASS, a passenger accelerated service system, using biometric kiosk for frequent travelers. So far, 130,000 users have used that system. At Ben Gurion Airport in Tel Aviv a similar system processes 80,000 frequent travelers per month through Tel Aviv Airport reliably and efficiently, and at great service to the passenger, and improved security. Later this year, 90,000 Palestinians will cross the land border into Israel using combined hand geometry and face recognition to expedite arrival into Israel for their daily work.

So we encourage looking at what has already been implemented and is able to be moved quickly for Registered Traveler, and look at the difference between operational and reference biometrics within the context of a Registered Traveler program.

Turning to the issue the Chairman made earlier about the vulnerability of the credentials of law enforcement officers getting on planes with firearms. This issue relates closely to the issue of TWIC. You really need to prevent two vulnerabilities today. One is that the current credentials are very easily copied, or lost, or stolen. So there is a copying threat. Second of all, there is no tying of the individual to the credential using biometrics. Similarly to TWIC, there is mature technology available today that could improve the security of that situation considerably.

Finally, one of the criticisms of the biometric industry has been around standards. It is not accurate to say that there are no standards in the industry. There are many efforts that have been concluded and some that are about to be concluded that the industry has been collaborating with this Government and international governments on a multitude of standards.

So, in conclusion, I agree that biometrics are not a panacea for aviation security. Security is a function of the weakest link in the technologies and in the processes within which those technologies are used. However, credentials are currently a weak link in aviation security and biometrics have proven themselves when tied to secure credentials to not only improve security, but improve convenience. So I propose that we move quickly out of this multitude of testing phases that we are in and into implementation of this technology. Thank you.

Mr. MICA. Well, that concludes our witnesses. I wanted these two witnesses to testify while our Government witnesses were here so they could hear just what was conveyed to the panel here. We are almost three years since we passed the Transportation and Aviation Security bill. Section 106 said: "The Administrator shall establish pilot programs in no fewer than 20 airports to test." On May 3, you announced three of these airports, Mr. Verdery, is that correct?

Mr. VERDERY. I am not sure if it was three exactly, but that is when the announcement came out, yes. I think it was eight.

Mr. MICA. What has taken so long?

Mr. VERDERY. This is on the access control issue, correct?

Mr. MICA. Access control, adoption of any standard so that we can get some biometric IDs in place.

Mr. VERDERY. Well, this is a complicated issue on all things biometric, and we have seen this——

Mr. MICA. Well, we just had this witness say that for 15 years they have been doing it at nuclear plants, 97 percent are covered. It sounds like we are studying the thing to death. We could have covered 40 of our major airports for the money we have spent on studying. How far are we away, Mr. Norton, from adopting a standard? Can they not adopt some standard and put something in place with some flexibility for changes in the future?

Mr. NORTON. The standards are largely in place to support the issuance of a——

Mr. MICA. The standards are largely in place?

Mr. NORTON. To support the issuance of a Transportation Worker ID Card.

Mr. MICA. Mr. Pearce, let me borrow his, this is his pilot license. It is almost a joke. Here are access cards for National Airport and none of them have a biometric component. Mr. Pearce's does not even have a photo, as handsome as he is. We do not even have facial recognition. And I doubt that he still weighs 175 pounds.

[Laughter.]

Mr. MICA. You cannot imagine how frustrating this is. Now DHS has basically assumed responsibility for developing the biometric standard; is that correct, Mr. Assistant Secretary?

Mr. VERDERY. Again, I just want to make sure that we are talking about the same issue. Are you talking about the armed LEO access, is that the issue?

Mr. MICA. Anything. Anything.

Mr. VERDERY. Well, we have responsibility for biometric standards for programs that fall within our Department, of which there are many.

Mr. MICA. OK. US-VISIT, Registered Traveler is going to be part of it. But again, just some standard identification. Until you move, until somebody in the Federal Government moves on adopting a standard, we have no identification that has a biometric standard for our U.S. Marshals, for our Capitol Police, for anyone. No one will move until you move in adopting a standard, not to mention the State and local governments.

Mr. VERDERY. If I could, because I do not want the record to be incomplete, NIST is in charge for the Federal Government in terms of doing the research and setting broad standards for Federal Government purposes for biometrics, and they have been ongoing in that research and we feel that their research has been very useful to us in the various programs that we are trying to implement, whether it is VISIT or these other things we have mentioned. Within our Department, the Science and Technology Directorate is in charge of taking that broad research and applying it to particular problems we want to solve.

Mr. MICA. To standards, adopting a standard.

Mr. VERDERY. Yes.

Mr. MICA. How far are we from adopting some standards? Somebody in the Federal Government has to adopt some standard to move forward. Is this a little vendor competition that is just going around in circles, or what?

Mr. VERDERY. No. We are working both within and without the Department on the broad identification standards that might apply for a range of biometric uses. But as the witnesses testified, we have the expertise in place largely on the technology side to do programs that are fairly simple biometrically, on what the biometric will be on a card and how it will be read, whether it is a TWIC program, and RT program, US-VISIT, border crossing cards, and the like.

Mr. MICA. You have adopted that standard and said this is the standard?

Mr. VERDERY. It is not a single standard, but we have the ability to take the biometric, usually a fingerprint but it could be other things, and to have a reader that can read that. That is not really the difficult issue. The difficult issue is more the deployment of the machines, the vetting of the passengers or the persons to make sure they have had an accurate watchlist and terrorist check, and also to make sure that can be updated.

Mr. MICA. We have no watchlist, a consolidated watchlist, so I guess we do not have to worry about that, which is something else we asked for almost three years ago.

Again, it is very frustrating that we have some airports—San Francisco, it was testified, has had a program in place for years now. But as far as iris, as far as facial recognition, as far as fingerprint, someone has to adopt a standard. Most of the privacy questions and things like that someone also has to answer so these systems can be put in place. But right now we have 18,000 credentialed law enforcement officers, we have pilots, we have airport access badges, none of them have any standard or biometric provision that allows us to say that person in fact is who has the card or the badge or the license. When can we expect to have something in place?

Mr. VERDERY. Each one of the situations you mentioned is a slightly different part of a large puzzle of identification and the use of biometrics, and each one of them we are looking at to try to come up with a tailored solution that fits the audience that we are trying to address. So, for Registered Travelers and for LEOs, we have the pilots we have announced this month that will be going into place next month and we hope to have the results very quickly.

Mr. MICA. Almost three years later after we asked. Section 136 also says: "The Under Secretary of Transportation for Security shall recommend to airport operators, within six months after the date of enactment, commercially available measures to prevent access to secure airports." Where are we on that?

Mr. VERDERY. Well, that is the guidance that has gone out for the Access Control pilots that were announced last month at the eight airports, we have announced eight out of ten.

Mr. MICA. And it says further: "Review the effectiveness of biometric systems currently in use at several airports." And we put specifically in the law, two and a half years ago, including San Francisco International. Do you have a copy of the review of that?

Mr. VERDERY. I do not have it with me. I am not aware, sir, as to exactly how the San Francisco experience is being utilized in the eight pilots that have been announced. But, obviously, we are very——

Mr. MICA. This was not like some of these things "may" provide the use of biometric, this is shall do certain things within a certain period of time. Can you provide the Subcommittee with your review, as required by Section 136 of the law?

Mr. VERDERY. I would be happy to go back and see if that has been accomplished, and if not, see when that might be.

Mr. MICA. OK. Were you aware that this report was found in the Al Qaeda cave, this is just the testimony, but the security breeches at Federal agencies and airports, were you aware of that?

Mr. VERDERY. Yes.

Mr. MICA. You were.

Mr. VERDERY. And it obviously highlights the issue. Yes, it is a problem.

Mr. MICA. Well, again, we are nearly three years out. Do you have regular meetings with all of the other agencies? Really, DHS is going to set the standard. But do you have regular meetings with all the other Federal agencies that may be adopting identification cards or——

Mr. VERDERY. Yes. OMB chairs an interagency working group that brings all the relevant players to the table on broad biometric policy development.

Mr. MICA. If I ask FAA, because we have been after them to get some sort of a pilot's license that does not look like it comes out of a Cracker Jacks box, they are going to tell me they have been part of that and you have been part of that meeting.

Mr. VERDERY. I cannot speak for FAA. I would be surprised if they were not there. I know DOT has been involved, but I cannot speak for FAA today. I know we have been involved both at a departmental level and through our component parts, whether it is TSA——

Mr. MICA. Do you think that it would be important given this kind of a report?

Mr. VERDERY. Of course. Of course.

Mr. MICA. OK. Again, it is very frustrating that we have not had—I mean, with everybody so concerned about screening passengers, and we know that the folks we are dealing with have access to the same information that we are talking about publicly here today——

Mr. VERDERY. Sir, if I could. I understand the frustration and I just would point out that in the last months, especially since we have had the full integration of the Department, we have been extremely active on getting biometrics into play, whether it is the VISIT program as it initially was deployed and our expansion we have announced to the visa waiver countries coming up this fall, the RT pilot we have announced, the TWIC pilot we have announced, the coordination of the APIS and ident fingerprint systems, which is absolutely crucial to making sure law enforcement has access to that biometric of criminals and other folks that we are worried about, our work in international bodies. I can just tell you, we are taking this issue extremely seriously and are using the technology that our good witnesses have provided for us.

Mr. MICA. OK. Now this OMB group, are you aware of State and local participation in those types of meetings or evaluation of how you are going forward?

Mr. VERDERY. I am not aware that there is official State or local representation in that group. I believe it is a Federal standard-setting or policy-making body. But we would obviously need to work with them on things like HAZMAT drivers and other things where the States have a very legitimate role to play.

Mr. MICA. Law enforcement officials. We have 18,000 different credentialed people able to carry arms. As you saw, I got a copy from my sheriff of what it requires, and it is very little. And, again, I have no problem with people carrying weapons. I do not care if they carry bazookas on aircraft if they do not pose a threat. Mohammed Atta is a different story with a bazooka. But we would never know who he is given the type of technology or lack of standards or biometric edification cards.

Mr. MICA. And we completely agree that the LEO issue is a serious one, a very legitimate one. We hope that the pilots we are deploying, which will be mandatory, will provide us the kind of knowledge to go out on a more broad basis for LEO control. We completely agree that this is a serious issue that needs to be addressed.

Mr. MICA. Even pilots. Pilots are flying the planes, and we have also had some reports of attempts to secure uniforms and things of that sort. Well, again, it is very frustrating. We can talk about certain things in this open forum, but we are limited. I think you know the situation we are in and we need some attention to this.

Mr. DeFazio?

Mr. DEFAZIO. Thanks, Mr. Chairman. Mr. Verdery, if we could sort of start with the basics. First, I think we can all agree we would like to know who the people are who have access to secure areas at airports and that they do not have either a terrorist or criminal background. We can agree on that; is that correct?

Mr. VERDERY. Of course, yes.

Mr. DEFAZIO. OK. And not to get you in trouble here with a superior—well, I do not know the chain of command, actually I think you are in a different chain of command—but Admiral Stone was here a couple of months ago and I raised the concern that in Europe not only do they have a much more advanced identification background check system, but, as they told us in Great Britain, we do not think an intense background check is adequate security for people who have access to the air side of the airport. We check anything and everything that goes on and off there. Admiral Stone says they actually have access to bomb-making materials and other things right there in the airport. And of course I said, plastic explosives, sheet explosives? No, no. There is fuel and there is this and that. Well, pretty primitive stuff. So he says there is no need to screen what is coming on and off the air side of the airport, and admits that we are doing background checks that are basically cursory, that we are not doing even enhanced background checks. Do you think that is adequate?

Mr. VERDERY. Well, I was not here when Admiral Stone testified, so I was not privy to the back and forth. But we are in the same chain of command. He is the Acting Administrator of TSA and we both report to the Under Secretary.

Mr. DEFAZIO. OK. OK. Well they kind of "disappeared" him, but, OK, since he is acting.

Mr. VERDERY. Yes. We think that folks who have access to the secure areas do deserve background checks, they are getting background checks.

Mr. DEFAZIO. Enhanced background checks. I mean, a background check that is beyond you run it through NCIC and come back with a negative. You do not know whether that is really that person or anything really about him, but you did not get a positive on that name which they have assumed or actually possess.

Mr. VERDERY. Well, again, there are two issues. There is somebody who has a possible hit that you need to chase down the lead and that is why the whole mechanisms we are putting in place starting with the terrorist screening center and through our watchlisting efforts, our efforts to run screening points through our ONRE database or through National Targeting Center, all these efforts to enhance the screening of people, whether it is airport workers or——

Mr. DEFAZIO. Are we talking physical screening or screening——

Mr. VERDERY. I am talking about vetting of a background check. Our capabilities there are becoming greatly enhanced throughout this year.

Mr. DEFAZIO. But as we move forward, at long last, with the trusted traveler—now I guess we are calling it, what is it, it has got a different name now?

Mr. VERDERY. Registered Traveler.

Mr. DEFAZIO. Registered Traveler Program, good, which I am very supportive of. We are saying that registered travelers who have had an intensive background check and who will have a counterfeit-proof ID, perhaps biometric, still need to take everything they are carrying with them and go through security. But vendors, employees, who may or may not have had anything other than a cursory background check, probably have not, caterers who have access directly to the airplane, cleaners who have access directly to the airplane, and others who have had a much less intensive background check do not need to go through any security and the things they are carrying on or off or around do not need to be screened. How do we justify that to the traveling public? The people I saw at Detroit were wearing big, bulky coats and could easily have had Uzis under them, taken them around through security, they did not go through a metal detector, unlike all the passengers out there, and met a passenger in the airport or maybe they had an e-ticket in their pocket and they were going to get on the plane themselves. The one answer I got before was, "Well, Congressman, they are not getting on the planes." I said, "How do you know they are not getting on the planes?" "Well, they work there." "Well, yes, but maybe they only work there as a cover and they are going on the plane." How do we justify that to the American public, that all these hundreds of thousands of people—how many people, let us just take La Guardia, how many SIDA badges are there at La Guardia?

Mr. VERDERY. I do not have the number in front of me.

Mr. DEFAZIO. Could you get it? I think it is tens of thousands.

Mr. VERDERY. Sure. It would obviously be quite large, a busy airport.

Mr. DEFAZIO. Right. Right. So all the passengers over here in the long lines, and the flight attendants and the pilots are over here in the long lines, and all these other people—and I do not know whether La Guardia is an airport that requires screening or not because you have not been able to get me a list on what airports require screening of employees, vendors, and others—but we know on the air side we are not screening them, for sure, because that is policy. Do you think this is a good plan? We did find box cutters concealed on the planes that were grounded after 9–11 that had not been used that were somehow smuggled onto the planes.

Mr. VERDERY. I think as we put together the RT program, the pilots, we are trying to come up with a plan that will attract the traveler and that——

Mr. DEFAZIO. Oh, no, we agree on RT. There is no problem with that. But what I am saying is the registered traveler has an intense background check, I am moving a little fast for you, but they are going to have a very intense background check, they are going to have a counterfeit-proof ID, but still everything they are carrying and they themselves have to go through screening, as poor as it is, or as good as it is. I think the screener is good, the equipment sucks, but that is a different issue not for today. We have all these other hundreds of thousands of people who have less intense background checks, who have various IDs that are not even uniform across the United States, going either around to the backside of the airport and the airplanes or right around security and into the airport at an unknown number of airports. Now what I am saying is, is that not a problem?

Mr. VERDERY. Well the question here is what level of background check is appropriate for folks who have access to the SIDA. That is something that we are going to be able to have better capabilities for once these mechanisms for vetting folks and the TSC is stood up. So this is not related to the RT issue except for people who happen to maybe see the folks who are not going through screening. They are separate issues, obviously, under the same rubric of aviation security.

Mr. DEFAZIO. Yes. But if I might, the fact is that enhanced background checks are available; you can buy them, they are out there. And we have decided as a Federal policy to not require enhanced background checks of all these hundreds of thousands, millions of people nationally who have access. We know who the pilots are, we know who the flight attendants are, we do not really know who the passengers are. That causes me a lot of concern. Does the technology exist, since we cannot decide on a biometric, to print and require re-badging everybody with a uniform national badge at least? Could we do that? Could we require that?

Mr. VERDERY. Would technology allow that? Sure.

Mr. DEFAZIO. Sure. Why do we not do that? Why do we have this multiplicity of badges across the country? We have already heard biometric is being used and deployed, but we have not been able to get there. But absent that, would it not at least give us some level of confidence if we re-badged all these millions of people who have access to airports and airplanes on a daily basis with a uniform ID nationally? It would not cost much, I would not think, and it might be worth it.

Mr. VERDERY. I think it would cost quite a bit. But as we look at——

Mr. DEFAZIO. How much do you think? I am just curious. Do you think it would cost less than one hijacked airplane being used as a weapon?

Mr. VERDERY. That is a very tough comparison to make. But I would err on the side of safety.

Mr. DEFAZIO. Well, then, I think we ought to re-badge people. I think in fact the Committee, as I recall, had some discussion two years ago that we were going to re-badge everybody. That we were going to pull all the badges and we were going to re-badge everybody. I guess maybe we did that but we did not do it with a uniform badge.

Mr. VERDERY. When we looked at the broad sweep of transportation security needs in the country, and part of that is screening people who have access to sensitive transportation facilities, whether they are airports or other types of facilities, the fact is that the airport workers have a better security regime now than most. And so, as we are looking at prioritization of resources——

Mr. DEFAZIO. Well what does that mean? Could you explain that? When we witnessed the people at Detroit Airport who pulled out one of those IDs, wore a big coat and carried a bag, and walked through without even going through a metal detector, is that better?

Mr. VERDERY. Compared to most of our ports and other transportation facilities, it is better.k

Mr. DEFAZIO. Wow. You have drawn a pretty low standard here. I mean, yes, it is better than walking out the door, but it is not exactly what I would consider to be——

Mr. VERDERY. And that is the point of the TWIC program, is to enhance security across the broad swath of——

Mr. DEFAZIO. The TWIC which we have not been able to develop in two years, and we have not figured out what it is going to look like or how we are going to do it, when we are going to do it. That's great.

Just one last question, Mr. Chairman, I know I am badgering the witness. But Mr. Huddart, tell me, is this rocket science? If we are doing biometrics at all in nuclear plants, why could we not apply it more broadly?

Mr. HUDDART. I am not sure, sir.

Mr. DEFAZIO. Could you give me a guess? Does it fail a lot? Is it undependable? Has there been big problems with it?

Mr. HUDDART. There are many applications that I have outlined where biometric technology has been very successful. All biometrics are different. Some have different fits for different applications. I personally think the issue of standards has been blown up way beyond what it is. For example, if I have a key to open a door at San Francisco Airport, there is no standard that exists that that key has to open a door in LAX or La Guardia. Why are we holding biometrics to somewhat of a different threshold and we are trying to create a perfect system when, in fact, it is taking the extra time to create a perfect system when we have deployments all over the world that can be looked at that have happened today. How did the Department of Energy do this?

Mr. DEFAZIO. They mandated it, I believe.

Mr. HUDDART. There are many access control standards that biometric devices——

Mr. DEFAZIO. One last question. Did they, and I have been to nuclear plants and I have seen the system but I cannot think back whether they all used the same system, did they mandate a standard system, or did they just mandate that it had to be a biometric system with people's backgrounds verified?

Mr. HUDDART. The Department of Energy conducted a study and consolidated, to my knowledge, all of the facilities around one particular biometric that met their needs the best.kl

Mr. DEFAZIO. OK. Great. Thank you. Thank you, Mr. Chairman.

Mr. MICA. I thank the gentleman. Mr. Ehlers?

Mr. EHLERS. Thank you, Mr. Chairman. Let me just get into some of the more technical aspects I am interested in. To what extent the Under Secretary of Science and Technology at DHS has been involved in this issue of biometrics, and to what extent have their scientists been consulted on this issue?

Mr. VERDERY. They have been working quite closely with the various program offices, whether it is the TSA programs that have been the topic today, or other programs like US-VISIT and other things. So we have a pretty good team effort between S&T and the programs at NBTS. They have essentially been given the lead by the Secretary to be the brains for the future, to take our grant programs and solicit proposals from the private sector, to understand better enhancements and the like. So I think we have been working quite well with them. Obviously, we are a new department with hiring up an standing up new facilities and the like, and there is obviously room for improvement, but I think it has been a pretty productive relationship to date.

Mr. EHLERS. And what role has the National Institute of Standards and Technology played in your effort?

Mr. VERDERY. The most notable part of this was they were required under the entry-exit laws that Congress passed to develop and certify the standard for the entry-exit system for biometrics. That was done early in 2003 through a report from NIST to the Attorney General and I believe the Secretary of State, and they chose the ten print-two print combination that we are using for US-VISIT. So that was probably the highlight of their involvement. But they have also been quite engaged at the technical level, working with our program offices to ascertain how the systems are going to work as they build up entries in the database. One particular example I am well aware of is, there is concern that the US-VISIT database will eventually grow too large in a two print form and will not be able to sustain the good six second response time we are working now because it will have too many false positives. We are really in unchartered waters building a database that is going to eventually be this big. So we have been working with NIST to try to figure out ways to extend the life of that database. Eventually, we may have to migrate to a different style of fingerprint capture.

Mr. EHLERS. So are you saying that in the biometric systems you have settled on using prints instead of other physical characteristics?

Mr. VERDERY. For the US-VISIT system, the base architecture is the two print capture at the visa issuance stage, a two print verification at the port of entry, yes, along with a digital photograph, as opposed to other biometrics.

Mr. EHLERS. But I am talking about if you go to a general system, do you not want something that yields fewer false positives?

Mr. VERDERY. Well the fingerprint system has worked quite well. The number of false positives is very low to date. The number is in the less than 1 percent range. And so it has actually worked quite well as opposed to other things that have been on the table, such as facial recognition which has a much higher false positive rate. So we need to continue to monitor how the system stands up as it grows larger, but we have been happy with it to date.

Mr. EHLERS. What about iris identification, what is the false positive rate there?

Mr. VERDERY. I am not aware—I will have to get back to you on the specific figures from NIST. I think the bigger concern with the iris is not the false positive but the difficulty of capture, especially in an action environment like a port of entry. With the lighting and the like, it is not easy to get a good enough picture to acquire an iris that could be matched quickly.

Mr. EHLERS. I find that a little hard to believe. And I believe the false positive rate is negative on that case. I mean, 1 percent is still pretty big; that is one out of every hundred people.

Mr. VERDERY. It is less than 1 percent. I think I have it in here somewhere, but it is quite low.

Mr. EHLERS. OK. I am just trying to get at what is the hold up, why are we not moving faster on this. You have heard the frustration from my two colleagues about the lack of progress. And you are saying the science is understood, the technology is there, and then the question is, what is the hold up? What is the problem? Because it seems to me that is the most difficult part. The implementation is relatively easy, but trying to decide on the best system and make sure that it works well is the most difficult part, and I understand you are saying that is done.

Mr. VERDERY. Well, no. I think that is why in these environments like the Registered Traveler, like TWIC, like LEO credentials, we are looking at different types of access points, whether it is what type of biometric it would be, how it is collected, and those kinds of things. But that is really only half the puzzle. The implementation part is quite a challenge. We are talking about multiple millions of people in the TWIC environment eventually who would have a need for that kind of card. How do you get the card to them? How do you acquire the biometric? What kind of updates do you need? Do you have everyone come in at once? I mean, these are very difficult questions. The US-VISIT program has been a great success, I think everyone would agree, but it also was funded I believe at $330 million for this year. It is an expensive, but effective, program. And the other programs that we are working through are much lower dollar. We are trying to make sure we have both the biometric technology in place and also a clear plan for implementation before we charge forward with a one-size-fits-all when we are talking about literally millions of potential customers.

Mr. EHLERS. OK. I guess I view it somewhat differently. It seems to me once you do the science and technology, then the implementation, although troublesome and complex, is relatively easy. Adopting a standard, deciding on the system to use is the tough part and the rest of it is mechanics. And I agree, if you need money, well, then, you ought to ask us for more money for the implementation.

Mr. VERDERY. The implementation is very difficult. Another issue we are working on is the border crossing card, which has a biometric built in for Mexican citizens who travel frequently across the border. It has a biometric but it has no mechanism for it to be read while a person is in a car without getting out. So we are looking at building in radio frequency technology into the card so that the person does not have to leave their vehicle and can still have the biometric information read and vetted as they cross. But getting that biometric in place on six million border crossing cards is a very difficult venture when these cards are good for ten years. Do you make everyone go back in and be retrofitted at the same time?

Do we phase it in? There are a lot of difficult implementation questions. And that is just one biometric issue.

Mr. EHLERS. Yes. And I would worry about using RFI technology because it is too easily tampered with. So I would not put a great deal of faith in that system.

I would yield back, Mr. Chairman.

Mr. MICA. Thank you. Mr. Honda?

Mr. HONDA. Thank you, Mr. Chairman. I appreciate having this hearing. Mr. Huddart had made a comment about the 20 pilot programs that were incorporated into the authorization bill. And I need some clarification. I thought I heard you say that portion had retarded the progress of the deployment of the pilot programs. You said that you started with eight now, but, from what I understood, if the pilot programs were not there, you could have been at 45 out of the 82 top airports. Could you clarify that for me please.

Mr. HUDDART. Sure. When I say that it may have slowed down, the company I work for talks to a lot of airport managers who have an interest in adopting biometrics, and since 9–11 several airports have actually implemented biometrics but they have done so somewhat reluctantly, and may other airports have decided to wait until there is clear direction that comes from the Government with regard to any standards or recommendations around the technologies. So the fact that the Government has come out and said we are going to test biometrics at airports has lead to a wait and see attitude on behalf of the airports and has slowed many airports from actually implementing.

Mr. HONDA. The purpose of having a pilot is to test the biometrics, but also to establish a process by which each airport, understanding their own characteristics, can look at the selection of over-the-counter newly emerging technologies to be prescribed to their own particular characteristics. That was the intent of the pilots.

So to TSA, my question is, there seems to be a lag in execution of programs. I share the frustration of my colleagues, and they have been here a hell of a lot longer than I have and I have caught up to them. And I think that their frustration is that they have seen a history of tragedies prior to 9–11 in trying to get security placed into airports, and then after 9–11 we put together a bill that directs our agencies to put together, and working with airports and airlines, to come up with strategies around each airport. And I agree that we should not have one-size-fits-all; that is stupid. But in the San Jose Airport, we put together a blue ribbon task force, in conjunction with those in technology, airlines, commercial, transport, those who run the airport facilities, looking at validation of individuals, validation of the property, and coming up with processes that could be standardized in terms of studying how you would apply technologies. Have you read this, Mr. Verdery?

Mr. VERDERY. I have not seen that report, but I would be very interested in seeing it.

Mr. HONDA. This was submitted to the Department of Transportation prior to the switch. We were told that this is something they needed to have in order to understand how to do it. You have plenty of people backing up this study. We have recommendations in looking at every aspect of airport security, including taking care of the issue of privacy and civil liberties. I would like a response from you within at least three weeks, if you can, because I hear we will get back to you but I do not hear a time line for which we can expect a response. So I would like a response on how this would fit, how this would facilitate the deployment of the pilot programs.

Mr. HONDA. And quite frankly, $8 million for eight project just escapes me because it is such a small amount of investment in a very critical arena that our entire economy and our homeland security centers around. On top of that, Mr. Chairman, I am not sure what the strategy for TSA is right now. It seems like it is disjointed and pretty much like what our intelligence community was prior to 9–11, that you had different stacks and none of them are talking to each other. It does not seem like there is any kind of communication or any concise overall strategy. And I would like some response on the suggestion what date you can get back to us on your response to this study and how it applies to deployment of biometrics.

Mr. VERDERY. Well, sir, when I said I had not seen it, I mean personally, although I would like to. I would hope that the folks within TSA, both in the national office working on these types of programs, and also at the San Jose Airport directly, would have seen this. And if they have not, they should. And I will make sure that they have.

Mr. HONDA. I am sorry, the director of what?

Mr. VERDERY. The San Jose Airport FSD should have seen your report, and I hope that he or she has.

Mr. HONDA. It is a he, and I agree, he should have.

Mr. VERDERY. Right. So I will make sure that happens if it has not happened already. In terms of your comments more broadly about TSA, as the supervising entity for TSA, we feel good with where they are. It is a very challenging mission. But we think we have got very good leadership over there and we, with the leadership of the Under Secretary Asa Hutchinson, we are working to harmonize the various pieces of the puzzle. As you mentioned, there are a number of different programs but they do fit together. They have come up with a layered security approach and these programs, whether it is the screener work, the air marshals, the LEOs, the RT project, all these things work together in harmony to minimize the chance that we are going to have an aviation security incident. Is there improvement that could be done to coordinate those functions? Of course. But I think we have seen a good improvement in that coordination since the Department stood up, and, perhaps as importantly, the interaction between TSA and the other parts of our Department, because they work together and have to work together on lots of things, especially with customs and border protection, on port of entry issues, and airport security issues, and with our investigative arm at IS to make sure that law enforcement incidents are investigated and handled properly. So, room for improvement? Of course. But we think we have had a good start.

Mr. HONDA. Through the Chair, Mr. Chairman, your job is not only to monitor TSA, but also to deploy what we are talking about—biometrics and the kind of technology we have there for airport security. Is that correct?

Mr. VERDERY. One of our responsibilities, yes.

Mr. HONDA. So airline security is one of your main concerns?

Mr. VERDERY. Definitely.

Mr. HONDA. And the deployment and the application of technology, including biometrics, is part of that. I do not get a sense that there is a coordinated or a thoughtful approach in looking at each airport and finding out how they are going to be doing it. I read the report on the selection of the eight airports. And I do not know all of them, but I do not get a sense that they are varied in their characteristics. So what is it that you expect out of these pilots that is going to help deploy the technology across the board? Is it the technology, or is it a process by which they will go through in order to understand how they are going to secure their airport?

Mr. VERDERY. Perhaps it is both. If you look at the list of the eight airports and the exact pilots that are being funded, it is a range of sizes of airports, mixes of types of travelers, and what the pilot going to do, whether it is attempting to secure the perimeter of the airport, particular parts of the airport, passengers, workers. We are trying to see different aspects of the puzzle and then stitch them together in something that could be deployed more widely, funding allowing.

Mr. HONDA. And that was the basis upon granting the pilot status?

Mr. VERDERY. That is the goal of the pilots.

Mr. HONDA. But was that a requirement in order to grant them the status of a pilot, that they had done that already, or are they going to start from scratch with the grant?

Mr. VERDERY. I am not sure, honestly, Congressman, as to what the broader characteristics were they were required to demonstrate. To be a pilot site, they had to submit forms talking about what the goals of their particular pilot were, and how they were going to work with TSA to provide that information back to us, and a number of other factors. But I am not sure exactly what they were required to do. I would be happy to get the submission form to you.

Mr. HONDA. I am not trying to beat up on you. I would like to see the evaluation document that is relative to San Jose Airport. It seems like this document is exactly what it is that you are talking about in terms of process by which you evaluate a site for validation of not only equipment and movement of passengers, but also of those who work there, including biometric. So can you tell me when you can have a response with the analysis of San Jose Airport and why it was not selected, and your response on what you think this report in its application to other airports would be?

Mr. VERDERY. I would need to double back to the TSA folks and understand exactly how the criteria selection were determined, what the application for the airport was. How about if I promise to get back to you ASAP with a time frame on when we can get together and provide more information?

Mr. HONDA. I did not get a date. You said ASAP.

Mr. VERDERY. I think we could get back to you within just a couple of days as to when we could get back with a more formal presentation as to the exact questions that you raised.

Mr. HONDA. So in two days we can get a time line from you by which——

Mr. VERDERY. That sounds reasonable to me.

Mr. HONDA. And then a response on the study, what do you think, how long would it take?

Mr. VERDERY. A couple of weeks, I will say.

Mr. HONDA. Two weeks for that. OK. Thanks, Mr. Chairman.

Mr. MICA. I thank the gentleman. Mr. Shuster?

Mr. SHUSTER. Thank you, Mr. Chairman. Mr. Verdery, I guess you feel the frustration over here of I think everybody on this Committee. I have been here three years now and we have been hearing the same thing over and over again about a number of different programs through TSA and now through Homeland Security. It is very frustrating. And, God forbid, that we have another terrorist incident, but if it happens, we are all going to get fired because we are not doing our job and I do not think we are doing it fast enough.

I am not quite sure now—I thought at first when I started to hear you speak that this was a standard problem and that is why we are not moving forward, then later on I hear you saying about implementation—but is it an implementation problem, or a standard problem, or both?

Mr. VERDERY. Is there a particular program you are talking about, or more broadly?

Mr. SHUSTER. I am talking about either the airport employee program or the Registered Traveler Program, either one of those two. Is it a standard problem or—using biometrics I guess is what I am asking.

Mr. VERDERY. For the Registered Traveler, it is more of an issue of selecting appropriate airports where we think that we can entice the traveler to actually want to use the program. Are there benefits that are going to make it interesting and useful for the person to go through the biometric and the background check. Certain airports are going to have better access to the way we can structure the checkpoints so that there is a benefit to them. Certain airports may have other benefits outside the security realm that would be enticing to a traveler. We need to work with the airlines to set up those checkpoints.

Mr. SHUSTER. So we can implement that today if we find out what is going to motivate a person to be in that program, is that what you are saying?

Mr. VERDERY. I think we are scheduled to deploy those pilots within a month or so. We are looking at five different pilot airports. Those have not been announced yet. We are working with the airports.

Mr. SHUSTER. So we have a standard for that? We can do that today if we had to?

Mr. VERDERY. I do not think they are going to be cookie cutter pilots Different pilots may have different enticements or different structures, depending on the particular airport.

Mr. SHUSTER. OK. I do not believe we are going to have a problem enticing travelers to do it. From what I hear from the business traveler, they are willing to do anything to get out of the line—$100, $200, do somersaults if they have to. So if the President ordered the TSA to have a Registered Traveler Program in place by the end of this month, or say the end of this year, do we have a standard in place to be able to move forward with that?

Mr. VERDERY. Yes. We do not need the President to order it. We are going to have it in place at these pilot sites I think the middle of next month or end of next month.

Mr. SHUSTER. OK. So we have a standard?

Mr. VERDERY. We have the biometric captured.

Mr. SHUSTER. You use biometrics?

Mr. VERDERY. Yes.

Mr. SHUSTER. OK. We have a standard. So, now going back to the airport employee situation, do we have a standard in place that we can use biometrics to be able to identify people? Is that correct?

Mr. VERDERY. No. That is why we are doing the pilots. We are trying to ascertain which biometric is going to work best in an active environment like an airport security system. That is why—I mean, I can walk through the different pilots that we are looking at.

Mr. SHUSTER. No, you do not have to do that. I think I got a pretty good understanding, and that is what I think Mr. Honda was talking about. San Francisco, I do not know if that was San Jose or San Francisco, but it has been in place, is my understanding. Are you utilizing that ten year or twelve year history to decide which standard is best?

Mr. VERDERY. Well, they are not one of the pilots because they have a good system in place. But we are using the knowledge that they——

Mr. SHUSTER. That answered my question. They have a good system in place. Why are we not modelling ourselves after a good system in place instead of trying to develop—and it is not just you, I hear this all the time in the Federal Government. We are always looking for something new when we have San Francisco that has something and it is working. We can go to the Israelis to their air system, it works. The INS I understand, somebody said here today, the INS has a biometrics in place. Our nuclear facilities are all utilizing biometrics. It is very frustrating to keep studying and studying. Let us put this in place. Let us get it out there. Because if something terrible happens, as I said, we are all going to get fired, and we all should get fired because we are not putting something in place. We have got to go out there and we have got to put it in place. We all know that it is not going to be perfect. We can see that if you look at the CAPs system, they have been developing that for years and I think I just heard they are not going to utilize it because it is not going to work the way we thought it would. And if we put something in place today, the technology is going to be new six months from now or a year from now. So we have got to move forward. We have got to put something out there and get the system moving into place.

Mr. VERDERY. But it is just not the case that we can take one successful program, whether it is San Francisco or somewhere else, and just replicate it everywhere. San Francisco, as I understand it, uses a hand geometry system which apparently works great for them, which we encourage. But if we are looking at a system where we want to build a check against a terrorist database, hand geometry does us no good. Our systems are based on fingerprints, names, dates of birth, other pieces of information. The biometric is the fingerprint as the base. So a hand geometry does no good to find a terrorist working at that airport. So we need to understand, is that trade-off worth it? And so that is what we are looking at.

Mr. SHUSTER. But right now we have nothing. We do not have anything in place. That is my point. Let us move forward with something. Let us get something in place. We just keep talking and studying about it.

Let me move to Mr. Huddart. What is your level of confidence that if we said to you set up 40 airports, can the folks in your industry do that now? Can we implement those things that will work?

Mr. HUDDART. Absolutely. The industry does it everyday in banks, in hospitals, in schools——

Mr. SHUSTER. McDonalds.

Mr. HUDDART. And McDonalds, yes.

Mr. SHUSTER. Mr. Norton, what is your view on that. Is that something we can put into place?

Mr. NORTON. Yes, it is. I would like to clarify. The comparison was made as to whether or not one technology could be used over another. Fingerprints and face recognition, iris recognition, for that matter, can be used to see if you have multiple identities and multiple enrollments. Hand geometry and a number of other capabilities can be used in an operational environment to make sure that you know who you are dealing with. So there are different uses of biometrics for different purposes.

I think it is important to understand what we have and what we really need in order to get these solutions out there. And what we have, and I think it has come up here today at the hearing, is mature technologies that are there to support the mission, whether it is screening people for enrollment, or screening people in an airport environment where they are trying to get access to a facility. We know they work within an airport footprint. Those analogues are there. We know they are supported by standards. And it is important that we perhaps define what we mean by standards. We mean the technical descriptions that enable these technologies to be deployed. And we know that we have a pretty good concept of operations about how these applications may work.

What we really need is then to move forward to the phase of defining what the requirements are, establishing a government and industry policy that is appropriate for the circumstance, and then moving forward with the processes of funding and technology selection. We see that these activities are underway in this area. I think we can put a lot of the technology questions behind us. And as we move forward now to some of the stickier issues on industry and government policy and the requirements definition, we can get over that phase quickly because the technologies are there and they are mature.

Mr. SHUSTER. One final question, Mr. Secretary, on the Registered Traveler Program. Are we including in the development of this program, these pilots, people that are travelers, business travelers? It is my understanding that there has not been a whole of participation from the business community in developing this pilot.

It seems to me that if we are developing a product, first and foremost, security has to take precedence, but if we want to attract them, we need to be talking to the traveler to say what will attract you. Are we doing that? Are we doing marketing research in that?

Mr. VERDERY. The business traveler along with the LEO is the target audience of these pilots because that is who we understand is most likely to want to use this program and are the people we are trying to assist as well as leveraging our own resources. But TSA is reaching out to the stakeholders here—the airlines via their frequent flyer clubs and the like, that is kind of the crowd we are looking at. And we need to work with the airlines because they may be able to offer things to people to induce them into the program, whether it is frequent flyer miles, or access to the lounge, these kinds of amenities that might be an inducement as well. So that is who we are working with and that is our target audience.

Mr. SHUSTER. We are working with the airlines or are we working with the traveler themselves?

Mr. VERDERY. With the airlines and with the travel industry. I am not aware of the particular meetings they are having on a day to day basis, but that is the target audience and the people we are trying to leverage.

Mr. SHUSTER. OK. Thank you very much.

Mr. MICA. Thank you. Ms. Johnson?

Mrs. JOHNSON. Thank you very much, Mr. Chairman, and thank you and the Ranking Member for having this hearing. I apologize for being late. I had to preside at another hearing prior to coming. Since 9–11 we have hired thousands of primarily new trained screeners, placed hundreds of air marshals on flights, required advance manifests and increased inspections of air cargo, armed pilots, and secured cockpit doors, and passengers still have to practically undress to come through security—without shoes, without belts, without anything but underwear and a little cover. I am really concerned about how it is possible to be discriminatory with the techniques and technology we are talking about now.

My congressional district has DFW, Love Field, and two other airports in it. And DFW has 52 million passengers each year and it is really the pillar of our economic growth there. We want to provide safety and security and certainly efficient passenger processing while we preserve privacy, which is almost gone with the process now. If we had just 1 percent of bad prints, that is 52,000, and that would mean standing in lines longer almost than we have now. I am really concerned about the cost of failure. You have that many people that is failing, then what is the perfection level of the biometric system, and as it relates to privacy and security? Give me an explanation as to how you think this is going to increase security.

Mr. VERDERY. Ma'am, if you are speaking about the Registered Traveler Pilot Programs or the eventual Registered Traveler Program we might have in place, this would be a voluntary system. So the privacy impact is something that the traveler would assume, but we would have robust privacy procedures in place on the use of the biometric and the like. It would not be a mandatory program. TSA, you may know, recently hired a Chief Privacy Officer, who is now on board, reporting up to our departmental Chief Privacy Officer, who has been quite aggressive in these areas and is an integral part of our policy-making process. But we feel comfortable that the biometric part of a Registered Traveler Program would have a very low false positive match. The US-VISIT system, which is a similar type of fingerprint verification, has had an extremely low error rate, far less than 1 percent. So we are confident that we are not going to have logistical problems with the program in terms of making sure people are who they say they are. There will be certain cases where there is a false positive or a false match and we need to have people on the ground who can resolve those. And that is part of the implementation strategy that is necessary.

But in general, we feel very good about the privacy protections that can be put in place for RT and we will be working those aggressively. In fact, just as a related matter, the Department this week received the first ever so-call adequacy finding from the European Union for our privacy protections for airline passenger data. This is being used by Customs and Border Protection on incoming flights. But it is the first time that a foreign government has ever been certified by the European Union, who has extremely tough privacy laws, for our privacy procedures and redress mechanisms that are in place for Customs and Border Protection. And that will be a very useful thing for securing the international travel to your airport and others.

Mrs. JOHNSON. You know, 35 percent of the residents in my area were born in other countries. And the stories that I hear from what they have to go through with profiling is kind of hard to take. How do we explain to many of them what this new technology will do to either keep them from being so highly profiled, or will it add more to it?

Mr. VERDERY. Well, of course, TSA, under our current operations, does not profile a person's race, gender, or other characteristics. It is not a factor in whether somebody is singled out for secondary screening. As you know, most people who are sent to secondary is either due to an alarm or due to the mechanism by which they bought their ticket, which has nothing to do with them, it is just a travel pattern. The beauty of the biometrics, though, is that it essentially makes it a personal evaluation—is this person who he or she says that they are. It has nothing to do with their race, or height, or gender, or anything. It is a one person versus many check. And it allows us to tailor our programs to look at the person. And so to the extent that there is any residual profiling out there, which, again, we do not support and do not believe is happening, the biometric usage, whether it is in TSA procedures or others, should minimize that. It is one of the beauties of the technology.

Mrs. JOHNSON. Thank you very much, Mr. Chairman.

Mr. MICA. I thank the gentlelady. Mr. Pearce?

Mr. PEARCE. Thank you, Mr. Chairman. During all this process, Mr. Verdery, this process of trying to get to the next step, you have had industry and airlines and everyone involved in the process?

Mr. VERDERY. Very much so, yes.

Mr. PEARCE. Mr. Huddart, the International Biometric Industry Association is a fairly small industry, biometrics is a fairly small industry. Do you know most of the people in that industry?

Mr. HUDDART. A good portion. We have about 25 members, leaders in the industry, who are members of our Association.

Mr. PEARCE. But is there a large industry outside your knowledge base?

Mr. HUDDART. I would say not generally, no.

Mr. PEARCE. Not generally. Do you know the biometric industry representatives who were sitting in the collaborative process or trying to get us through the roadblocks?

Mr. HUDDART. Personally, not. That does not mean to say that there was not, but personally I do not have knowledge of that.

Mr. HUDDART. You had representatives from the biometric industry in there in trying to figure your way through this logjam of problems?

Mr. VERDERY. The folks within our Department, the program agency managers and the likes, are working closely with the potential vendors, with other people who have good ideas. If we have not made a connection between this particular Association and——

Mr. PEARCE. No, no. But I mean you did have biometric industry representatives in your meetings?

Mr. VERDERY. Yes. I mean, they are brainstorming, they are soliciting ideas. There obviously are rules on the contracting process between the official interaction. But yes, they are, and should be, part of our brainstorming and idea process. And I am a couple of layers above the program heads, but I would want the biometrics people who are experts in this to come in to see us and to give us an idea of what is possible and what the roadblocks are.

Mr. PEARCE. What is your educational background?

Mr. VERDERY. My background? College and legal degrees.

Mr. PEARCE. And your boss?

Mr. VERDERY. Under Secretary Hutchinson, I believe the same, college and legal, then he served in Congress.

Mr. PEARCE. How about the two layers below you?

Mr. VERDERY. Well, there are a lot of layers below me, but probably a wide variety. We have scientists——

Mr. PEARCE. Just the two layers, the two people right below you in your department, what is their background?

Mr. VERDERY. Well, within my particular office, we would be talking about policy analysts who might be lawyers, might be former Federal law enforcement agents. It is a variety of people that work directly for me.

Mr. PEARCE. So no airline experience, no biometric experience. You have got legal experience, you have got no operational experience.

Mr. VERDERY. No. We do have folks, I am not going to claim that somebody in my immediate office is an expert on biometrics, although we feel like we are getting that way. But within our apparatus of policy-making, there are people who are quite expert on this issue. The US-VISIT office has a Chief Strategist who is very experienced in biometrics.

Mr. PEARCE. Are there people with operational backgrounds at a level that can unplug the logjam, or are the decision-makers that have the logjam in place the people with legal backgrounds, legal educations?

Mr. VERDERY. No. We have a wide variety of people who are making decisions, some of them are lawyers, some of them are not, some of them are technologists, some of them are accountants, we have TSAs.

Mr. PEARCE. So the logjam originates up and down through the management spectrum. You do not have someone at the top who can say break the logjam who can make a decision. That is what the Chairman was saying, when are we going to make a decision. You have said that is a legal person, and you are a legal person. Tell me, Mr. Verdery, we have got this logjam sitting in place, what is going to change in the foreseeable future? The logjam is about cost, the logjam is a problem of implementation—I am reading from your statement—the logjam is vetting passengers, the logjam is civil liberties. What in the near future is going to change? Which one of those problems is going to go away? We have not broken the logjam because those problems exist. The technology is here, Mr. Huddart has said so, you have said so. The standards are here, Mr. Norton has said so. The standards and the technology are here, so we have got a logjam going because of these implementation problems. And which of those problems is going to go away to cause us to suddenly be able to do what we should be doing?

Mr. VERDERY. These types of problems never vanish, it is just going to take an incredible amount of hard work from people all the way from the Secretary on down to the program managers to fight through these. And I think you have seen from our remarks today, we are moving extremely aggressively. Again, just in the past month we have announced deployment on Registered Traveler, on LEO, on TWIC, on VISIT. There is an incredible amount of activity underway. Is it complete? Of course no. But this is not stuff that is sitting in a dusty room somewhere. This is stuff that is being deployed out in the field.

Mr. PEARCE. In my opinion sitting here, and I see my time has expired, but in my opinion, it is like you are trying to hit a hole in one. You are going to play golf by hitting a hole in one. I am sorry, a hole in one happens occasionally. Most of us have to hit the dadgummed ball and it is going to go over here, it is going to go over here, and you progressively get closer to the pin.

Now we have decided that we are not going to give anyone anything except these pilot certificates because we cannot vet passengers, and because we have got six million people coming across the Mexico border that we cannot read their stuff. Even if we had one group and we begin to correct that and say, OK, now we have got that problem solved, and, yes, it is going to have to be tweaked, but we have got it solved. And the Registered Traveler, we do not know exactly how we are going to pay for it, but I suspect if we asked the people to raise their hands, they would pay for the dadgummed thing themselves, and those who did not could stay in the long lines. If nothing else, we could begin to do some implementation.

But we have got a logjam and I do not really see anything that is going to change. You are going to have 20 pilot programs. And Mr Shuster said it well, why do you not use the one that is working? Well, it works out there but it might not work here. So you are going to have 20 pilot programs. What is going to make them work anywhere except where they are working in the pilot program? It seems to me that we have got enough people with legal expertise but not enough people with operational backgrounds that just know that you have to start somewhere and you have to improve on what you get and that you are not going to get a hole in one. Even when you hit the ball off the tee, when you finally pull the trigger on something, it is not going to go in the hole.

You are going to have spent years vetting and trying to get the process worked down and getting the nerve to make a decision, and the cost is going to remain the same, the obstacles of implementation are going to remain the same, and the civil liberties are going to remain the same, and the vetting problem is going to remain the same. They do not disappear over time. They do not just cure themselves. So somebody is going to have to have the courage to do something, sometime, somewhere and maybe get the ball moving forward.

I hope that these pilot projects that you are doing right now are that move. Frankly, being frozen in the headlights does not go away, and so what we generally do is we move from one frozen position to the next. I think that we are going to get the pilot projects and we are going to be frozen in place as to what to do with them because I do not think we are really integrating people. I do not think we have gone to Israel to study and, if they have already solved the problem, why we have not imported it back to here instead of creating it all from the ground up. You can make any response you would like. Thank you, Mr. Chairman.

Mr. VERDERY. Sir, if I could. Again, I think your description of not waiting for the perfect shot or perfect deployment, I used to play golf before I took this job and I catch the analogy. But for instance, in the US-VISIT, Congress had mandated an entry-exit system going back all the way to 1996 and it did not happen because people could never agree on kind of what the grand plan was going to be. It was, well, there is no point in deploying it here if you have not done it here. And so nothing ever happened. Well, this Department started on I believe it was March 1, and April 29th, it was actually my second in the job, Secretary Ridge announced, no, this is the plan, we are going to have entry-exit at the end of this year. It is not going to be universal because we are doing the first phase. Airports and seaports, it is in place, we beat the deadline, we beat Congress' mandate, we got it in place, we are finding bad guys every single day. And now we have to move on to the next building blocks—land borders, visa waiver countries, the exit scenarios which are incredibly important. But we took I will not call it a baby step because it was a huge step, but it part of the answer.

These other issues that we are working on, it is a similar thing. We understand we are not going to be able to deploy a TWIC program to every one, I think it is 21 million is the number I heard, of workers in one fell swoop. We have got to figure out a deployment plan and make sure it works.

So I think there is leadership coming down from the top from the Secretary and Under Secretary Hutchinson to make these programs a reality. And I feel confident—I am not going to put a time frame sitting here—but there is progress underway. And if there was a logjam, the logjam has been broken.

Mr. PEARCE. Thank you, Mr. Chairman.

Mr. MICA. Mr. Verdery, how long have you been Assistant Secretary?

Mr. VERDERY. I was confirmed last June.

Mr. MICA. Well, as you can tell, there is a sense of great frustration here. Mr. DeFazio and I, he is the Ranking Member, we have decided we are going to, if necessary, change the law. So we are going to hold a markup the first week when we get back after Memorial Day, and instead of "may" we are going to have "shall," and then we are going to define what we want you to do very specifically. So you can go back and tell Secretary Hutchinson, Mr. Ridge, anybody above or below, we are going to try to define what we want you to do and get it done as quickly as possible. Everybody is frustrated with this. The test airports that you are doing, they are not all biometric. I see a number of video surveillance projects. They are not all biometric, are they?

Mr. VERDERY. I believe you are correct. I need to find the list again. But I think that is right, some are and some are not.

Mr. MICA. So we are not even talking about eight biometrics. And again, this has all been done somewhere. If we could get you all to set some standards for the different types of functions, or adopt the nuclear plant standard, some standard, the rest we believe will begin to fall in place. What is really frustrating about the test airports, and I talked about the test airports before, I do not think one of these test airports has an integrated in-line check baggage system. Tampa does not, T.G. Green State does not, Southwest Florida does not, Savannah does not, Newark does not, Minneapolis, Miami does not, Boise does not. Just from common sense, can we do—we have eight to fourteen, depending on who we ask, integrated in-line check baggage systems across the country. Can we not have one airport where we have a biometric system in place for access control with all the whistles and bells in one place? Would that not make sense to have sort of a model, maybe East Coast, one in the center of the United States, so people could go and at least look at the technology? San Francisco is not 100 percent but they have not done that. But just from somebody's thinking, and we talked about this before, have one place where we can show the latest technology operational for a choice of airports that wanted this, would that not make sense?

Mr. VERDERY. It might. I was not part of the grant selection process here in determining which particular airports were going to be awarded these.

Mr. MICA. Somebody had better get in charge of things over there and start thinking about this before we have a disaster. And the way we spend this money is just so frustrating. You cannot imagine how frustrating. I am surprised you have been around as long as you have and somebody has not taken control of these programs and made some sense out of them.

Mr. DeFazio, did you have anything else?

Mr. DEFAZIO. Thank you, Mr. Chairman. Mr. Chairman, you have expressed I believe the sentiments not only of yourself and myself but many other members of the Committee who could not be here because of other obligations. And I expect we will have a unanimous vote when we move toward a realistic mandate for implementing some of these systems. Mr. Verdery, not to lay it all on you. You are not the first person to be in the hot seat here. We went from Mr. Magaw to Admiral Loy, and they all managed to squirt out and move on somewhere else, and Admiral Stone, they do not let him come up anymore, or he does not want to, I am not sure which. But you are here today and you are the guy. But you have got to understand that we are not doing this to be petty. We just feel an extraordinary sense of urgency about these things. And I realize there is a huge range of threats and when I start to think about the wider scope beyond this Committee's jurisdiction, I start to get bogged down a little bit, too. OK, well, gee, what are we going to do on the ports, what are we going to do on trains, another Subcommittee I am on. But for aviation I think we can do better without an extraordinary expense.

Mr. Huddart, I just want to explore a little more about who is using—you know, we can look at the nuclear plants and say, well, that involves a few hundred people at each plant, there are not that many of them, and this is not really a very big model to say the technology has been working for fifteen years and we could implement it elsewhere. Could you give us other examples? What is the broadest example? What does DoD do, for instance?

Mr. HUDDART. There are many DoD. In fact, the Air Force has used biometrics extensively for base security. So, for example, at Scott Air Force Base there are turnstiles. That when you arrive to Scott Air Force Base, it is an unattended application, they use biometrics to get through those turnstiles, to verify your identity. I do not know how many users there are but it must be several thousand.

Mr. DEFAZIO. And that takes care of the follow-on thing that we are still trying to figure out a way to deal with—piggybacking, they call it?

Mr. HUDDART. Yes. The turnstiles prevent piggybacking.

Mr. DEFAZIO. Only one person can get in the turnstile, or the turnstile is surveyed remotely to see.

Mr. HUDDART. That is correct. That is one example. As I mentioned, San Francisco is quite a large application. There are 15,000 daily users who probably use the system each four, six, eight times a day.

Mr. DEFAZIO. And what biometrics are they using?

Mr. HUDDART. Hand geometry biometrics, the same as the nuclear industry.

Mr. DEFAZIO. OK. And what is the false positive? I mean, is it tested regularly? Do people try and defeat it?

Mr. HUDDART. The Department of Energy has done extensive testing on different biometrics. There has been several different tests done within the Government. There are really two attributes when you are designing a system of a biometric you have got to look at. One is the false positives, one is the false negatives. Both errors that different biometrics can make. In the case of that particular biometric, it is in the range of .2 to 1 percent depending on the particular application.

Mr. DEFAZIO. That was .2 to 1?

Mr. HUDDART. Yes, .2 to 1 percent, depending on the application.

Mr. DEFAZIO. What is the delay time? I plunk down my hand, how long does it take?

Mr. HUDDART. The total transaction time is about three seconds, in that range.

Mr. DEFAZIO. Three seconds. That is pretty good.

Mr. HUDDART. And enrollment time is generally less than a minute.

Mr. DEFAZIO. OK. Anybody else have anything they did not get a chance to say that they would like to say on my time?

Mr. VERDERY. If I could sir, just as a conclusion. We share your sense of urgency. These programs are being developed as rapidly as we can develop them within the funds that we have.

Mr. DEFAZIO. Aha. The key point. And you are not allowed to ask for more funds, I know that. So, OK.

Mr. VERDERY. I can tell you though that this travel facilitation issue along with the security side is something that we are focused on quite a bit. Within my office, we are just a part of the puzzle here, of course, but Under Secretary Hutchinson would tell you the same thing, that he is anxious to get both the RT program in place to try to help alleviate some of the crowding at the airports and facilitate the business travel we know is so important, and the LEO issue, as we talked about, incredibly important, we have got to show progress there, and we think that this will be a step in the right direction.

So I just would not want the record to be closed without sharing your sense of urgency both within TSA and up the food chain.

Mr. DEFAZIO. I appreciate that. I do not denigrate the motivations of anybody in these matters. And I know you would not be working there if you did not take it seriously. But the key thing, and I have said it to Admiral Stone, and I have said it to agency people all the time, and, unfortunately, I know the constraints and it is not just under Republican Administration, you are not allowed to come and tell us you need more money. But the fact is we are not going to get the security that I think the Chairman and I want for the American traveling public on the cheap. When you are up there, and I have had this debate on safety with the airlines and others, I have yet to sit next to someone who says, I really do not mind there is a terrorist on the plane because I got a really cheap ticket, so it is OK with me. No. There is no one up there saying that. So we have got to work our way through this. And I think, under the Chairman's leadership, when we put forward a mandate, then you may well be able to pass the ball back to us and say, OK, we have got the mandate, here is what it is going to take, here is what it is going to cost. Ultimately, Congress has got the power of the purse. So that may be a way to get you out of the hot seat and pass the thing back to us. Thank you very much. Thank you, Mr. Chairman.

Mr. MICA. Thank you. I am even more frustrated by reading where they have given these grants. And this is not going to get us any closer to a solution. You have got video surveillance technology in at least two of these; then you have got biometric radio frequency technology to control access vehicles, that is testing wireless capability to transmit data, that is not going to solve our problem; you have got one iris and I see two fingerprint technology readers. All this stuff has been done. If you all would just go back and say we have got to make a decision, we have got to adopt a standard for identification for law enforcement people, for airport workers, for other Federal workers, and the rest will fall into place. I strongly believe that. But these tests, all of this has been done, maybe with the exception of the wireless transmission. And if you want to go ahead and look at that. But the technology has been developed. Standards are there. Adopt something that can be used by these people. The airports would reissue these badges. They lost, somebody just testified, what, one airport lost 400 badges.

Mr. DEFAZIO. Mr. Chairman, if I could. On this wireless issue, are we not using the wireless on the Canadian border for people who transit the border frequently? I think we are.

Mr. VERDERY. The Nexus program on the northern border and Century for the southern border for vehicle access.

Mr. DEFAZIO. OK. So we have already got it working. I do not know why we need to test it.

Mr. MICA. I know. But I do not see anything. Somebody over there please make a decision and the rest will fall in place, we guarantee it. Local law enforcement will get a card and it will have the biometric requirement that you have and we will be able to tell, with some modicum of certainty, that that is that individual that has got a loaded weapon and going on an aircraft. Cost-benefit, there is no cost. You are not going to absorb it for local government. Just set a standard. Sit down with these people, and it is not a very big circle of people participating in this, just a handful. Of course, there are going to be some losers, some vendors. Again, I am wondering if it is just vendors keeping this stirred up so nobody wins a prize. But it does not even appear to be that.

If you all could move forward in some way. I talked to other agencies preliminary to this hearing, they are all waiting for DHS to make a decision before anybody else moves. No local law enforcement is going to move, no State agency is going to move, no other Federal agency is going to move until you make a decision. I know it is hard and somebody has to assume responsibility and go forward with this. But there is nothing here in any of these tests that is going to come up with anything new that I know of that you could not make a decision on a standard or a technology. And what kind of reader is acceptable? We have readers, do we not, guys, that read these things?

Mr. NORTON. Yes, sir.

Mr. HUDDART. Yes, sir.

Mr. MICA. We were over in Amsterdam and we saw that they have iris—was it iris that I failed? Yes, iris. We have tested them. They tested iris, Sue Myrick told me, more than a year ago at Charlotte Airport I believe. San Francisco. Just do something. OK? I just do not know what else to say. This is the most frustrating thing I have ever been involved with.

Now we are going to change the law and we are going to direct you to do something. We all agree. People are coming up to me, "Why can't they do something?" It is not a partisan issue; not a Republican, not a Democrat issue. And it is nothing against you guys. We love you. We wish you well. But somebody has got to make a decision so that we at least get something in place.

Mr. DeFazio has asked unanimous consent that the record be left open for a period of two weeks. Without objection, so ordered.

No other business to come before this Subcommittee, this hearing is adjourned.

[Whereupon, at 12:20 p.m., the committee was adjourned, to reconvene at the call of the Chair.]

International Biometric Industry Association

Statement of
Martin Huddart

Chairman of the Board
International Biometric Industry Association

Before the
Subcommittee on Aviation
Committee on Transportation and Infrastructure
U.S. House of Representatives

May 19, 2004

Mr. Chairman and members of the subcommittee, thank you for inviting the biometric industry to offer its views at this important proceeding. My name is Martin Huddart. I am the Vice President of Business Development for the Electronic Access and Biometric Groups at Ingersoll-Rand. Recognition Systems, Inc., a subsidiary of Ingersoll-Rand, is the developer and manufacturer of a hand geometry biometric system and also offers fingerprint biometric solutions.

I am also Chairman of the Board of Directors of the International Biometric Industry Association (IBIA), and I represent IBIA here today. IBIA was founded in 1998 and is headquartered in Washington, D.C. IBIA's members are leading developers, manufacturers, and integrators of the full range of biometric technologies.

Overview about Biometrics. Biometrics are technologies that automatically identify or verify the identity of an individual by measuring physiological or behavioral characteristics. This authentication of identity is accomplished by using computer technology in a noninvasive way to match patterns of live individuals in real time against enrolled records. Examples of the patterns used for biometric identification include those made from the image of a fingerprint, the geometry of the hand, and unique patterns in a person's iris, voice, signature, or face. It is important to know that most biometric applications do not store the actual image of the feature being measured. Instead, the measurements are converted into a biometric file which is generally encrypted. Without the key to unlock the encryption, a biometric file cannot be reverse engineered to determine a person's name, age, sex, race or any other personal information. Likewise, it cannot be abused to steal someone's identity. In short, biometrics, properly used, protect privacy.

Biometrics are the only technologies that offer an effective response to the need for automated personal authentication as an essential component of strong homeland security systems without sacrificing convenience. The U.S. government was an early adopter of biometrics, first using the technologies to control access to highly sensitive facilities such as nuclear power plants and weapons storage locations. Now, use of biometrics is expanding to protect networks against intrusion by hackers, to secure records from identity theft, to ensure that benefits are disbursed to lawful recipients, and – not least – to protect international borders.

Continuing Threats to Aviation. Government and private industry have recognized the need for systems of positive personal identification – specifically by deploying biometrics – since 9/11. It is now widely acknowledged that terrorism, and indeed all criminal activity, thrives in an atmosphere of anonymity and false identity. The crucial issue is balancing the necessity for positive identification with our desires for a free and open society. Freedom to travel, a treasured benefit in our democracy, is exploited and corrupted by those who would threaten all movement, all travel, thus creating the appearance of imminent danger in the attempt to impose fear on our population and cripple the economy. We need to deny them that opportunity without sacrificing our rights of travel.

Many efforts have been made since 9/11 to address the need for additional security through biometrics in the aviation environment. Most were well-intended and necessary initial steps to improve air travelers' security, but they have also been piecemeal, hurried, and reactive. Accordingly, this statement by IBIA addresses remaining gaps in aviation security that can be filled by biometrics in order to help create a well-designed and comprehensive deterrent against terrorism in the aviation sector. With proper care, IBIA's recommendations to improve aviation security can also be leveraged to create striking improvements in passenger convenience and airline productivity that will help revitalize the aviation industry and encourage expanded travel and tourism.

Required Upgrades of Employee Identification to Strengthen Physical Access Controls. On May 7, 2004 TSA issued a Request for Proposal (RFP) for the Transportation Worker Identification Credential (TWIC). The RFP is the result of extensive consultation with industry by TSA, and it serves as the central guideline for employee identification in order to strengthen physical access controls for air, sea, and land transport workers. The RPF makes clear that TSA has considered an end-to-end solution. First, biometrics will be collected and enrolled to establish the identity of transport workers. After a background check by TSA, transport workers will be issued a credential that will hold a biometric. Workers' biometrics will also be retained in computer systems for future re-issuance in cases of lost or stolen credentials. Finally the TWIC system will use the biometric stored on the credential to integrate identity management and access control in local systems at airports, seaports, rail, pipeline, trucking and mass transit facilities.

The TWIC identification system will add needed clarity to the current TSA regulation governing the security of sensitive areas of airports. The current regulation reads as follows:

> "(a) Secured area. Except as provided in paragraph (b) of this section, the measures for controlling entry to the secured area required under §1542.201(b)(1) must:
>
> (1) Ensure that only those individuals authorized to have unescorted access to the secured area are able to gain entry;"

Biometrics are not stipulated in this regulation but – as the TWIC RFP recognizes -- biometrics in fact are the only secure way to authorize personal access in sensitive areas of airports. Most airports currently address the current TSA regulation by requiring personnel to swipe a card through a reader and enter a personal identification number (PIN). This system has wholesale vulnerabilities. Cards authorize access not to persons but only to pieces of plastic that are subject to loss, theft, or copying. Recently, a Category X airport – which includes the largest U.S. airports -- admitted that its annual identification badge loss exceeded 400 per year, a very large number. By contrast, airport personnel enrolled in biometric systems cannot transfer their identity to someone else, and their biometric information cannot be borrowed and used by an unauthorized party. Moreover, advanced versions of biometric access control systems combine the technology with sophisticated software that can limit users to certain airport doorways at certain times, and can track who accesses which door at what time.

Hand geometry is in use, for example, in airports at San Francisco, Nevada, and Toledo. An additional 15 U.S. airports are conducting trials of hand geometry at single entry points. Fingerprint controls are in use at Little Rock, Arkansas and Chicago O'Hare. Iris technology has been deployed at Terminal 4, the international arrivals terminal, at JFK Airport in New York.

These are rare exceptions, however. Many other airports are delaying a decision to deploy biometrics until the completion of testing of "new and emerging security technologies." These tests, being conducted at 20 airports, were mandated by the Aviation Security Act of November 2001. The law also provides that the Under Secretary for Transportation Security "shall review the effectiveness of biometrics systems currently in use at several United States airports."

The test process appears to be preoccupied with "new and emerging" technologies at the expense of deployed, proven technologies. For example, none of the first eight airports in the test uses hand geometry, which has proved its effectiveness in airport deployments that predate 9/11. Thus far the test managers have not reviewed the effectiveness of any operational biometric system already in place at an airport. It is not clear why. The test managers themselves say that in the end they will not recommend any biometric over any other and airports will be able to choose among proven biometric systems, yet the conclusion of the current test process appears to be at least a year away.

This long delay is unnecessary. Any of several biometrics that have proved their effectiveness in years of airport deployment could be approved for deployment today at all U.S. airports. Other biometrics could be approved later when tests demonstrate their effectiveness. Moreover, at least part of the funds being expended in the overly prolonged test process could be used for actual biometric deployment. My own calculation is that the money allocated to the test process could have retrofitted approximately 45 of the top 200 airports with biometrics.

The TWIC system has been structured to accommodate multiple biometrics, and it requires no more delay in the "testing" process. It is long past time to strengthen physical access control of personnel at airports by deploying biometrics properly for personal identification.

Improvements Needed to Identify Air Travelers. In the same way that TSA has adopted a comprehensive approach to airport security through TWIC, TSA must also adopt a comprehensive and holistic "registered traveler program." Post-9/11 security requirements have made air travel less convenient but only minimally safer. Deploying biometrics to positively identify travelers using a voluntary system could improve air travel security and convenience.

On April 5, Rear Admiral David M. Stone, Acting Administrator of the Transportation Security Administration (TSA), announced that the agency is seeking responses from the private sector to an RFP for a Registered Traveler (RT) Pilot Program that will begin in select airports in late June.

The RT Pilot will use biometrics to enhance security and efficiency. It is intended to create an information technology system that will fully integrate biometric identification with the results of security assessments to ensure fast, secure, and reliable personal identification and reliable measures of security status at airport checkpoints. The RT Pilot Program will ask volunteers to submit information, including biometrics, necessary for TSA to determine eligibility. The biometric information will be used to verify identity and in conjunction with a security assessment will allow passengers to pass through an expedited airport security screening process. All volunteers will continue to undergo basic physical screening procedures.

Biometric technologies have demonstrated their ability to eliminate bottlenecks in secured processing environments. The clearest example of this capability is in border control. Biometrics have been used in the most sensitive national security applications to routinely admit pre-registered border crossers. One of the best examples is the Israeli-Palestinian border project. Palestinians daily enter and exit Israel in order to conduct their business, visit families, and work in Israel. The 40,000 workers arriving daily from Gaza need to enter Israel within a three-hour period and exit at the end of the working day. A manual check would require hundreds of persons to man security checkpoints without a guarantee of reliability. By using biometrics, people entering or exiting Israel can be verified or rejected within seconds.

Palestinians wishing to enter Israel are issued a highly secure smart card after first enrolling in the system, receiving clearance that they have no previous terrorist or criminal record and that they have not previously enrolled in the system under an alias. The smart card holds substantial information, including biometric templates and personal and security data.

A Palestinian wishing to legitimately enter or exit Israel at a border crossing checkpoint presents a smart card at a biometric kiosk then places his or her hand on a reader, is biometrically verified as claimed, and after being cleared, proceeds through an open gate. Biometrics thus allow Israel to automatically verify a person's identity in the shortest possible time, in a user-friendly way, while maintaining a high level of security.

The Israeli-Palestinian border project is a prototype of how the TSA Registered Traveler program might work to ease air travel bottlenecks and simultaneously strengthen security.

Essential Changes in Credentialing of Law Enforcement Officers Carrying Weapons. Verifying the identity of authorized law enforcement officers who carry firearms onto planes is not a new issue but it remains a matter of real vulnerability. The issue arose well before 9/11 but has gained greater salience since then.

Credentials presented by law enforcement officers differ greatly but they are as unreliable as drivers' licenses to verify identity and they suffer from the same inherent problems of insecurity. Law enforcement officers' credentials typically consist of documents containing descriptions, photographs, and/or signatures. It is thoroughly insecure to try to verify personal identity by relying upon descriptions, photos, or signatures that are neither intended nor designed to be an integral component of an automated biometric identification system. A process this insecure is an open invitation to criminal and terrorist deception.

The General Services Administration and some state governments have begun to issue credentials (badges, drivers' licenses, and entitlement benefit cards, for example) that include an encrypted biometric template, but most government identification documents currently include no biometric. A digital photo standing alone, commonly used on identification credentials, is a wholly inadequate means of personal identification. Without standardized biometric authentication, attempts to use photos to achieve a valid 1:1 match is equivalent to the "garbage in, garbage out" aphorism often suggested by computer programmers.

To be acceptable, a law enforcement officer's credential presented to a TSA official must prove that the bearer is who he or she claims to be. For the same reason that biometrics are essential to authenticate the personal identity of transport workers and airline passengers, biometrics are required to prove the identity of law enforcement officers. Using a biometric 1:1 match to affirm the validity of the credentials held by a law enforcement officer is indispensable to helping deter the use of stolen or forged documents by criminals or terrorists posing as law enforcement officers.

Standards for Biometric Implementation at Airports and for Interoperability. Operational standards to implement biometrics at airports are being defined by both the TWIC program and the Registered Traveler program. Both will need to be further refined as the systems are deployed. In addition, the US-VISIT program is setting standards for air passenger security. These programs will help define operational guidelines to protect airports and air travelers. In addition, the biometric industry is hard at work to define standards for interoperability.

Notably, the biometric industry and government have worked together to develop a set of rules about how biometrics are to be integrated into computer operating systems. This is an exceptionally important advancement for several reasons:

- It accommodates multiple biometrics.
- It allows the quick adoption of new biometric technologies as they are deployed.
- It permits the rapid exchange of information for record checks.
- It enables users to voluntarily share biometric information that has been acquired by other sources, such as employers, airlines, and government agencies.

It is sometimes said that "the biometric industry has no standards." The statement is not accurate, but there is confusion about the alphabet soup of biometric standards initiatives under way domestically and internationally.

In fact considerable progress has been achieved. The BioAPI Consortium, a voluntary initiative driven by the U.S. biometric industry, is far along in balloting a base interoperability standard for biometrics through the International Organization for Standardization (ISO). The BioAPI, or Biometric Application Programming Interface, already serves as the cornerstone for interoperability in the Federal government. GSA, TSA, and the Department of Defense's Biometric Management Office require compliance with BioAPI as a condition of Federal government procurement of biometrics.

Beyond the initiative to achieve a base interoperability standard, standards initiatives in particular applications are proceeding through the American National Standards Institute (ANSI), through ISO working groups, and through the UN-recognized International Civil Aviation Organization (ICAO). These initiatives are developing standards for border crossing documentation – meaning biometrically-enabled passports and visas – for multi-modal biometric interoperability, for smart card and biometric interoperability, and for privacy and template security. The National Biometric Security Project, scheduled to present testimony at the May 19 hearing of the Aviation Subcommittee, is playing a central and vital role in all of these initiatives.

Conclusion. The need to deploy biometrics to help ensure aviation security is no longer a matter of real debate. Rather, the urgent task is to implement a coherent, holistic plan to deploy biometric technologies with all deliberate speed in the applications in which biometrics can clearly strengthen the security of airports and air travel. They include using biometrics:

- To control physical access to sensitive airport facilities.
- To identify airport and airline employees.
- To verify the identity of air travelers.
- To protect against unauthorized carrying of firearms on planes.

IBIA stands ready to support legislation and other initiatives by the Subcommittee on Aviation to advance toward these goals.

SUBCOMMITTEE ON AVIATION
HEARING ON THE USE OF BIOMETRICS TO IMPROVE AVIATION SECURITY
WEDNESDAY, MAY 19, 2004 @ 10A.M.

Thank you Mr. Chairman.
I commend you for your leadership on this matter and welcome our witnesses here this morning.

As September 11, 2001 clearly displayed, terrorists adapt, lie, forge documents, and attempt to disguise their intentions. Since that time we have made noteworthy strides in strengthening aviation security.

We have hired tens of thousands of highly trained screeners, placed hundreds of air marshals on flights, required advanced manifests, increased inspections of air cargo, armed pilots, and secured cockpit doors.

The economy of the Dallas-Fort Worth region is heavily dependent on a healthy aviation industry.

My congressional district is home to two commercial airports, once of which is Dallas-Fort Worth International Airport—the sixth largest airport in the world. It services 52 million passengers a year; employs thousands of my constituents; and serves as a pillar of economic growth for the region.

Providing safety, security, and efficient passenger processing while preserving privacy is of critical importance to my constituency; thus, I have followed the developments involving biometrics with keen interests and have a few concerns:

1.) **Research shows some biometric technologies are discriminatory**. Many people have fingers that simply do not print well. Even if people with bad prints represent one percent (1%) of D/FW International's annual passenger population, this would mean massive inconvenience and suspicion for that minority. A one percent (1%) error rate would mean fifty-two thousand (52,000) errors, each of which translates into lost aviation resources resulting from false leads.

2.) **The cost of failure is high.** If you lose a credit card, you can cancel it and get a new one. If you lose a biometric, you've lost it for life. Any biometric system must be built to the highest level of data security, privacy must be designed into them from the beginning, and encompass system-wide architecture to prevent compromise by corrupt or deceitful agents within the organization.

Mr. Chairman, my congressional district exists within a region where thirty-five percent (35%) of the residents were born in another country or are the children of foreign born. These individuals are a vital component to our social culture, and it is important that I be able to explain to them what implications this technology will have on them.

Again, I want to welcome our witnesses this morning, and I look forward to there testimony.

National Biometric Security Project

U.S. House of Representatives

Committee on Transportation and Infrastructure

Subcommittee on Aviation

Using Biometrics to Improve Aviation Security

Testimony by

Richard E. Norton
Executive Vice President

May 19, 2004

National Biometric Security Project
Suite 390 South
601 13th Street, N.W.
Washington, DC 20004
(202) 347-9788

About NBSP and Biometric Research

The National Biometric Security Project (NBSP) is a nonprofit research foundation that was established in 2002 to improve national security by developing and deploying advanced biometric technologies. The specific mission of the NBSP is to provide the Federal Government with research and development capabilities that will give the civilian government and private sector critical infrastructure the tools needed to secure facilities and sensitive data from compromise and intrusion by unauthorized people.

NBSP, headquartered in Washington, DC, is primarily funded under an appropriation by Congress under the Biometrics in National Security program. NBSP maintains a major testing and research center in Morgantown, West Virginia. Our primary activities are focused in five major areas:

1. Conducting applied research to determine security requirements, and developing solutions that can be implemented under rigorous operational conditions.
2. Establishing training and education programs to develop U.S. expertise in biometric technology.
3. Testing and evaluating biometric products to determine if they can fulfill operational requirements.
4. Maintaining and distributing information about biometric products and how they can be used to meet security needs.
5. Establishing standards that will simplify the tasks of selecting, implementing and operating biometric-based security solutions in real-world environments.

Biometrics and Aviation Security: Lessons Learned

Biometrics have been in use at airport facilities for over a decade. Federal aviation regulations have long supported the installation of biometric systems to guard sensitive areas of airports against intrusion, and a significant body of information exists on how airports have managed the processes of deploying, operating and administering biometric-based security solutions.

The most prominent example is offered by San Francisco International Airport, which uses over 170 biometric devices to protect ramps and jetways against unauthorized access. Employees who are enrolled in the system use a combination of a personal information number or ID card and their biometric (in this case, hand geometry) to gain admittance to secure areas. According to airport officials the system operates reliably under demanding conditions. Major airports in New York, Los Angeles, Chicago, Miami and Newark have also used biometrics to varying degrees to upgrade security.

As noted above, Federal regulations clearly supported the use of biometrics to enhance security levels at airports since the early 1990s. However, since the regulations only offered biometrics as one of several options for complying with security requirements, they have not been deployed as part of a mandatory program. Following the events of September 11, however, Congress took decisive steps to accelerate the installation of biometric-based security solutions and mandated trials of biometrics to secure the airport footprint. Under the Airport Access Control Pilot Program, the Transportation Security

Administration (TSA) will evaluate how biometric devices operate when they are installed at 20 airports throughout the U.S. beginning later this year. Metrics from that program should provide key indicators on the effectiveness of a wide array of biometrics. Among the most critical of these will be "usability" measurements that compare ease of operation against the business and security imperatives of the aviation industry.

Large-Scale Deployment: The Next Challenge

Putting biometrics into service at a particular airport is a question that has already been largely answered. Based on the years of experience acquired at San Francisco, NBSP expects that the existing body of data combined with the results of the TSA trials should clearly settle any lingering doubts about whether biometrics can significantly improve security controls at any given commercial airport. In situations where these systems are used daily by a trained group of users, biometrics have proven to be a particularly effective means of discouraging attempts to gain unauthorized access to facilities.

The challenge will be to settle on a model that will allow biometrics to be used effectively on a national basis by a broad group of transportation workers. Requiring those with regular access to a specific airport to enroll in a biometric security system presents no significant cost, administrative or procedural barriers. Expanding coverage to a national infrastructure level raises several major issues to be resolved:

- How to make biometric systems interoperable without mandating that a particular biometric solution be used across the board;
- How to ensure that people are not enrolling under an assumed identity or with multiple identities;
- How to implement a national system on a cost-effective basis; and
- How to guarantee that privacy requirements can be met and that data is adequately safeguarded against compromise and abuse.

The TSA has started to work on these issues under the auspices of the Transportation Worker Identity Card (TWIC) program. The TWIC concept calls for the issuance of a card to all transportation employees, including those who must have access to airport facilities in the performance of their duties. To date, TSA has established a concept of operations for the TWIC program, which calls for applicants to be pre-screened for multiple identities and criminal record through the use of fingerprint and face recognition biometrics. This is an essential process that can be met through the use of existing biometric technology.

Beyond the enrollment stage, the TWIC architecture will accommodate the use of other biometrics to perform the key task of controlling access in an operational environment. The design will permit an airport to incorporate its existing biometric door access systems within the TWIC design, or make use of different biometrics to solve specific operational requirements as they are identified. This ability to upgrade or change technologies seamlessly as new capabilities are developed is a necessary attribute of any well-designed security system.

Following its research on card types operational concepts, TSA is actively examining the existing infrastructure that is in place at transportation facilities nationwide, with a view to leveraging capabilities and resources that are already in place. NBSP agrees that this is a wise approach. Creating a standalone TWIC program from scratch would be prohibitively expensive, and local expertise will provide important insight on how the system can be installed without disrupting economically vital transportation systems.

Pre-Deployment Biometric Testing and Research Support

As these critical tasks move forward and biometric-based solutions come closer to full-scale implementation at airports, NBSP is actively leading a number of initiatives that should help ease problems with deployment. As a first step, NBSP is accelerating the development of standards that will enable biometric data to be accurately stored and accessed in a wide variety of systems. Working with the National Institute of Standards and Technology (NIST) and international standards organizations, NBSP has placed a top priority on assuring that interoperability problems are removed as a barrier to the broad adaptation of biometric technology.

Next, NBSP is equipped to handle the demands of a testing regime that can evaluate and certify the effectiveness of biometric products and solutions prior to field installation. NBSP laboratories are being designed to subject devices and integrated systems to stringent examinations that will determine if they can operate under a wide range of conditions. These tests will include evaluations of usability, durability, sensitivity to ambient environmental conditions, and performance against established requirements.

Finally, NBSP is building a cadre of trained biometrics professionals who can assist government and the private sector critical infrastructure to develop requirements for biometric systems and oversee their installation. Together with NBSP's database of information on biometric technologies and applications, the Project offers Federal, state, and local authorities an unprecedented capability that can help them speed up the adoption of practical, effective solutions that use biometrics to achieve new levels of security.

We appreciate this opportunity to testify before the Subcommittee on Aviation concerning this vital topic, and look forward to answering any questions you may have.

United States General Accounting Office

GAO

Testimony

Before the Subcommittee on Aviation, Committee on Transportation and Infrastructure, House of Representatives

For Release on Delivery
Expected at 10:00 a.m. EDT
Wednesday, May 19, 2004

AVIATION SECURITY

Challenges in Using Biometric Technologies

Statement of Keith A. Rhodes, Chief Technologist
Applied Research and Methods

GAO
Accountability * Integrity * Reliability

GAO-04-785T

May 19, 2004

GAO
Accountability-Integrity-Reliability

Highlights

Highlights of GAO-04-785T, a testimony before the Subcommittee on Aviation, Committee on Transportation and Infrastructure, House of Representatives

AVIATION SECURITY

Challenges in Using Biometric Technologies

Why GAO Did This Study

One of the primary functions of any security system is the control of people moving into or out of protected areas, such as physical buildings, information systems, and our national border. Technologies called biometrics can automate the identification of people by one or more of their distinct physical or behavioral characteristics. The term biometrics covers a wide range of technologies that can be used to verify identity by measuring and analyzing human characteristics—relying on attributes of the individual instead of things the individual may have or know. Since the September 11, 2001, terrorist attacks, laws have been passed that require a more extensive use of biometric technologies in the federal government.

In 2002, GAO conducted a technology assessment on the use of biometrics for border security. GAO was asked to testify about the issues that it raised in the report, the current state of the technology, and the application of biometrics to aviation security.

www.gao.gov/cgi-bin/getrpt?GAO-04-785T.

To view the full product, including the scope and methodology, click on the link above. For more information, contact Keith Rhodes at (202) 512-6412 or rhodesk@gao.gov.

What GAO Found

Biometric technologies are available today that can be used for aviation security. Biometric technologies vary in complexity, capabilities, and performance, and can be used to verify or establish a person's identity. Leading biometric technologies include facial recognition, fingerprint recognition, hand geometry, and iris recognition. The Federal Aviation Administration (FAA), and subsequently, the Department of Homeland Security (DHS) and the Transportation Security Administration (TSA), has been examining the use of biometrics for aviation security for several years. TSA has three current pilot projects that will study the use of biometrics to enhance aviation security: the Transportation Worker Identification Credential (TWIC), registered traveler, and an access control pilot program designed to secure sensitive areas of an airport.

It is important to bear in mind that effective security cannot be achieved by relying on technology alone. Technology and people must work together as part of an overall security process. Weaknesses in any of these areas diminish the effectiveness of the security process. The security process needs to account for limitations in biometric technology. For example, some people cannot enroll in a biometrics system because they lack the appropriate body part. Similarly, errors sometimes occur during matching operations. Exception processing that is not as good as biometric-based primary processing could be exploited as a security hole. Further, non-technological processes for enrollment are critical to the success of a biometrics-based identity management system. Before a person is granted a biometric credential, the issuing authority needs to assure itself that the person is eligible to receive such a credential.

We have found that three key considerations need to be addressed before a decision is made to design, develop, and implement biometrics into a security system:

1. Decisions must be made on how the technology will be used.
2. A detailed cost-benefit analysis must be conducted to determine that the benefits gained from a system outweigh the costs.
3. A trade-off analysis must be conducted between the increased security, which the use of biometrics would provide, and the effect on areas such as privacy and convenience.

Security concerns need to be balanced with practical cost and operational considerations as well as political and economic interests. A risk management approach can help federal agencies identify and address security concerns. To develop security systems with biometrics, the high-level goals of these systems need to be defined, and the concept of operations that will embody the people, process, and technologies required to achieve these goals needs to be developed. With these answers, the proper role of biometric technologies in aviation security can be determined.

Mr. Chairman and Members of the Subcommittee:

I appreciate the opportunity to participate in today's hearing on the use of biometrics for aviation security. The security of the U.S. commercial aviation system has been a long-standing concern. Following the September 11, 2001, terrorist attacks, virtually all aviation security responsibilities now reside within the Department of Homeland Security (DHS) and its Transportation Security Administration (TSA). These responsibilities include the conduct of passenger and baggage screening and overseeing security measures for airports, commercial aircraft, air cargo, and general aviation. DHS and TSA have undertaken several initiatives to improve aviation security. Some efforts, including those involving access control to secure areas of an airport and identifying travelers, include biometric technologies.

One of the primary functions of any security system is the control of people moving into or out of protected areas, such as physical buildings, information systems, and our national border. People are identified by three basic means: by something they know, something they have, or something they are. People and systems regularly use these means to identify people in everyday life. For example, members of a community routinely recognize one another by how they look or how their voices sound—by something they are. Automated teller machines (ATM) recognize customers from their presentation of a bank card—something they have—and their entering a personal identification number (PIN)—something they know. Using keys to enter a locked building is another example of using something you have. More secure systems may combine two or more of these approaches.

Technologies called biometrics can automate the identification of people by one or more of their distinct physical or behavioral characteristics—by something they are. The term biometrics covers a wide range of technologies that can be used to verify identity by measuring and analyzing human characteristics. Biometrics theoretically represent a more effective approach to security because each person's characteristics are thought to be distinct and, when compared with identification cards and passwords, are less easily lost, stolen, counterfeited, or otherwise compromised.

As requested, I will provide an overview of biometric technologies that are currently available, describe some of the current uses of these technologies, and discuss the issues and challenges associated with the implementation of biometrics. My testimony today is based on a body of work we completed in 2002 that examined the use of biometrics for border control. In that report, we discussed the maturity of several biometric technologies, the possible implementation of these technologies in current border control processes, and the policy implications and key considerations for using these technologies.[1] We also researched selected prior and current TSA and DHS biometrics initiatives and summarize them in this statement. We performed our work in accordance with generally accepted government auditing standards.

Biometric Technologies for Personal Identification

When used for personal identification, biometric technologies measure and analyze human physiological and behavioral characteristics. Identifying a person's physiological characteristics is based on direct measurement of a part of the body—fingertips, hand geometry, facial geometry, and eye retinas and irises. The corresponding biometric technologies are fingerprint recognition, hand geometry, and facial, retina, and iris recognition. Identifying behavioral characteristics is based on data derived from actions, such as speech and signature, the corresponding biometrics being speaker recognition and signature recognition. Unlike conventional identification methods that use something you have, such as an identification card to gain access to a building, or something you know, such as a password to log on to a computer system, these characteristics are integral to something you are.

How Biometric Technologies Work

Biometric technologies vary in complexity, capabilities, and performance, but all share several elements. Biometric identification systems are essentially pattern recognition systems. They use acquisition devices such as cameras and scanning devices to capture images, recordings, or measurements of an individual's characteristics and computer hardware and software to extract, encode, store, and compare these characteristics. Because the process is automated, biometric decision-making is generally very fast, in most cases taking only a few seconds in real time.

Depending on the application, biometric systems can be used in one of two modes: verification or identification. Verification—also called authentication—is used to verify a person's identity—that is, to authenticate that individuals are who they say they are. Identification is

[1] U.S. General Accounting Office, *Technology Assessment: Using Biometrics for Border Security*, GAO-03-174 (Washington, D.C.: Nov. 15, 2002).

used to establish a person's identity—that is, to determine who a person is. Although biometric technologies measure different characteristics in substantially different ways, all biometric systems start with an enrollment stage followed by a matching stage that can use either verification or identification.

Enrollment

In enrollment, a biometric system is trained to identify a specific person. The person first provides an identifier, such as an identity card. The biometric is linked to the identity specified on the identification document. He or she then presents the biometric (e.g., fingertips, hand, or iris) to an acquisition device. The distinctive features are located and one or more samples are extracted, encoded, and stored as a reference template for future comparisons. Depending on the technology, the biometric sample may be collected as an image, a recording, or a record of related dynamic measurements. How biometric systems extract features and encode and store information in the template is based on the system vendor's proprietary algorithms. Template size varies depending on the vendor and the technology. Templates can be stored remotely in a central database or within a biometric reader device itself; their small size also allows for storage on smart cards or tokens.

Minute changes in positioning, distance, pressure, environment, and other factors influence the generation of a template. Consequently, each time an individual's biometric data are captured, the new template is likely to be unique. Depending on the biometric system, a person may need to present biometric data several times in order to enroll. Either the reference template may then represent an amalgam of the captured data or several enrollment templates may be stored. The quality of the template or templates is critical in the overall success of the biometric application. Because biometric features can change over time, people may have to reenroll to update their reference template. Some technologies can update the reference template during matching operations.

The enrollment process also depends on the quality of the identifier the enrollee presents. The reference template is linked to the identity specified on the identification document. If the identification document does not specify the individual's true identity, the reference template will be linked to a false identity.

Verification

In verification systems, the step after enrollment is to verify that a person is who he or she claims to be (i.e., the person who enrolled). After the individual provides an identifier, the biometric is presented, which the biometric system captures, generating a trial template that is based on the vendor's algorithm. The system then compares the trial biometric template with this person's reference template, which was stored in the system during enrollment, to determine whether the individual's trial and stored templates match.

Verification is often referred to as 1:1 (one-to-one) matching. Verification systems can contain databases ranging from dozens to millions of enrolled templates but are always predicated on matching an individual's presented biometric against his or her reference template. Nearly all verification systems can render a match–no-match decision in less than a second. A system that requires employees to authenticate their claimed identities before granting them access to secure buildings or to computers is a verification application.

Identification

In identification systems, the step after enrollment is to identify who the person is. Unlike verification systems, no identifier is provided. To find a match, instead of locating and comparing the person's reference template against his or her presented biometric, the trial template is compared against the stored reference templates of all individuals enrolled in the system. Identification systems are referred to as 1:N (one-to-N, or one-to-many) matching because an individual's biometric is compared against multiple biometric templates in the system's database.

There are two types of identification systems: positive and negative. Positive identification systems are designed to ensure that an individual's biometric is enrolled in the database. The anticipated result of a search is a match. A typical positive identification system controls access to a secure building or secure computer by checking anyone who seeks access against a database of enrolled employees. The goal is to determine whether a person seeking access can be identified as having been enrolled in the system.

Negative identification systems are designed to ensure that a person's biometric information is not present in a database. The anticipated result of a search is a nonmatch. Comparing a person's biometric information against a database of all who are registered in a public benefits program, for example, can ensure that this person is not "double dipping" by using fraudulent documentation to register under multiple identities.

Another type of negative identification system is a watch list system. Such systems are designed to identify people on the watch list and alert authorities for appropriate action. For all other people, the system is to check that they are not on the watch list and allow them normal passage.

The people whose biometrics are in the database in these systems may not have provided them voluntarily. For instance, for a surveillance system, the biometric may be faces captured from mug shots provided by a law enforcement agency.

Matches Are Based on Threshold Settings

No match is ever perfect in either a verification or an identification system, because every time a biometric is captured, the template is likely to be unique. Therefore, biometric systems can be configured to make a match or no-match decision, based on a predefined number, referred to as a threshold, that establishes the acceptable degree of similarity between the trial template and the enrolled reference template. After the comparison, a score representing the degree of similarity is generated, and this score is compared to the threshold to make a match or no-match decision. Depending on the setting of the threshold in identification systems, sometimes several reference templates can be considered matches to the trial template, with the better scores corresponding to better matches.

Leading Biometric Technologies

A growing number of biometric technologies have been proposed over the past several years, but only in the past 5 years have the leading ones become more widely deployed. Some technologies are better suited to specific applications than others, and some are more acceptable to users. We describe seven leading biometric technologies:

- Facial Recognition
- Fingerprint Recognition
- Hand Geometry
- Iris Recognition
- Retina Recognition
- Signature Recognition
- Speaker Recognition

Facial Recognition

Facial recognition technology identifies people by analyzing features of the face that are not easily altered—the upper outlines of the eye sockets, the areas around the cheekbones, and the sides of the mouth. The technology is typically used to compare a live facial scan to a stored template, but it can also be used in comparing static images such as digitized passport photographs. Facial recognition can be used in both verification and identification systems. In addition, because facial images can be captured from video cameras, facial recognition is the only biometric that can be used for surveillance purposes.

Fingerprint Recognition

Fingerprint recognition is one of the best known and most widely used biometric technologies. Automated systems have been commercially available since the early 1970s, and at the time of our study, we found there were more than 75 fingerprint recognition technology companies. Until recently, fingerprint recognition was used primarily in law enforcement applications.

Fingerprint recognition technology extracts features from impressions made by the distinct ridges on the fingertips. The fingerprints can be either flat or rolled. A flat print captures only an impression of the central area between the fingertip and the first knuckle; a rolled print captures ridges on both sides of the finger.

An image of the fingerprint is captured by a scanner, enhanced, and converted into a template. Scanner technologies can be optical, silicon, or ultrasound technologies. Ultrasound, while potentially the most accurate, has not been demonstrated in widespread use. In 2002, we found that optical scanners were the most commonly used. During enhancement, "noise" caused by such things as dirt, cuts, scars, and creases or dry, wet, or worn fingerprints is reduced, and the definition of the ridges is enhanced. Approximately 80 percent of vendors base their algorithms on the extraction of minutiae points relating to breaks in the ridges of the fingertips. Other algorithms are based on extracting ridge patterns.

Hand Geometry

Hand geometry systems have been in use for almost 30 years for access control to facilities ranging from nuclear power plants to day care centers. Hand geometry technology takes 96 measurements of the hand, including the width, height, and length of the fingers; distances between joints; and shapes of the knuckles.

Hand geometry systems use an optical camera and light-emitting diodes with mirrors and reflectors to capture two orthogonal two-dimensional images of the back and sides of the hand. Although the basic shape of an individual's hand remains relatively stable over his or her lifetime, natural and environmental factors can cause slight changes. The shape and size of our hands are reasonably diverse, but are not highly distinctive. Thus, hand geometry is not suitable for performing identification matches.

Iris Recognition

Iris recognition technology is based on the distinctly colored ring surrounding the pupil of the eye. Made from elastic connective tissue, the iris is a very rich source of biometric data, having approximately 266 distinctive characteristics. These include the trabecular meshwork, a tissue that gives the appearance of dividing the iris radially, with striations, rings, furrows, a corona, and freckles. Iris recognition technology uses about 173 of these distinctive characteristics. These characteristics, which are formed during the 8th month of gestation, reportedly remain stable throughout a person's lifetime, except in cases of injury. Iris recognition can be used in both verification and identification systems.

Iris recognition systems use a small, high-quality camera to capture a black and white, high-resolution image of the iris. The systems then define the boundaries of the iris, establish a coordinate system over the iris, and define the zones for analysis within the coordinate system.

Retina Recognition

Retina recognition technology captures and analyzes the patterns of blood vessels on the thin nerve on the back of the eyeball that processes light entering through the pupil. Retinal patterns are highly distinctive traits. Every eye has its own totally unique pattern of blood vessels; even the eyes of identical twins are distinct. Although each pattern normally remains stable over a person's lifetime, it can be affected by diseases such as glaucoma, diabetes, high blood pressure, and autoimmune deficiency syndrome.

The fact that the retina is small, internal, and difficult to measure makes capturing its image more difficult than most biometric technologies. An individual must position the eye very close to the lens of the retina-scan device, gaze directly into the lens, and remain perfectly still while focusing on a revolving light while a small camera scans the retina through the pupil. Any movement can interfere with the process and can require restarting. Enrollment can easily take more than a minute.

Signature Recognition

Signature recognition authenticates identity by measuring handwritten signatures. The signature is treated as a series of movements that contain unique biometric data, such as personal rhythm, acceleration, and pressure flow. Unlike electronic signature capture, which treats the signature as a graphic image, signature recognition technology measures how the signature is signed.

In a signature recognition system, a person signs his or her name on a digitized graphics tablet or personal digital assistant. The system analyzes signature dynamics such as speed, relative speed, stroke order, stroke count, and pressure. The technology can also track each person's natural signature fluctuations over time. The signature dynamics information is encrypted and compressed into a template.

Speaker Recognition

Differences in how different people's voices sound result from a combination of physiological differences in the shape of vocal tracts and learned speaking habits. Speaker recognition technology uses these differences to discriminate between speakers.

During enrollment, speaker recognition systems capture samples of a person's speech by having him or her speak some predetermined information into a microphone a number of times. This information, known as a passphrase, can be a piece of information such as a name, birth month, birth city, or favorite color or a sequence of numbers. Text independent systems are also available that recognize a speaker without using a predefined phrase. This phrase is converted from analog to digital format, and the distinctive vocal characteristics, such as pitch, cadence, and tone, are extracted, and a speaker model is established. A template is then generated and stored for future comparisons.

Speaker recognition can be used to verify a person's claimed identity or to identify a particular person. It is often used where voice is the only available biometric identifier, such as telephone and call centers.

Accuracy of Biometric Technology

Biometrics is a young technology, having only recently reached the point at which basic matching performance can be acceptably deployed. It is necessary to analyze several metrics to determine the strengths and weaknesses of each technology and vendor for a given application.

The three key performance metrics are false match rate (FMR), false nonmatch rate (FNMR), and failure to enroll rate (FTER). A false match occurs when a system incorrectly matches an identity, and FMR is the probability of individuals being wrongly matched. In verification and positive identification systems, unauthorized people can be granted access to facilities or resources as the result of incorrect matches. In a negative identification system, the result of a false match may be to deny access. For example, if a new applicant to a public benefits program is falsely matched with a person previously enrolled in that program under another identity, the applicant may be denied access to benefits.

A false nonmatch occurs when a system rejects a valid identity, and FNMR is the probability of valid individuals being wrongly not matched. In verification and positive identification systems, people can be denied access to some facility or resource as the result of a system's failure to make a correct match. In negative identification systems, the result of a false nonmatch may be that a person is granted access to resources to which he or she should be denied. For example, if a person who has enrolled in a public benefits program under another identity is not correctly matched, he or she will succeed in gaining fraudulent access to benefits.

False matches may occur because there is a high degree of similarity between two individuals' characteristics. False nonmatches occur because there is not a sufficiently strong similarity between an individual's enrollment and trial templates, which could be caused by any number of conditions. For example, an individual's biometric data may have changed as a result of aging or injury. If biometric systems were perfect, both error rates would be zero. However, because biometric systems cannot identify individuals with 100 percent accuracy, a trade-off exists between the two.

False match and nonmatch rates are inversely related; they must, therefore, always be assessed in tandem, and acceptable risk levels must be balanced with the disadvantages of inconvenience. For example, in access control, perfect security would require denying access to everyone. Conversely, granting access to everyone would result in denying access to no one. Obviously, neither extreme is reasonable, and biometric systems must operate somewhere between the two.

For most applications, how much risk one is willing to tolerate is the overriding factor, which translates into determining the acceptable FMR. The greater the risk entailed by a false match, the lower the tolerable FMR. For example, an application that controlled access to a secure area would require that the FMR be set low, which would result in a high FNMR. However, an application that controlled access to a bank's ATM might have to sacrifice some degree of security and set a higher FMR (and hence a lower FNMR) to avoid the risk of irritating legitimate customers by wrongly rejecting them. As figure 1 shows, selecting a lower FMR increases the FNMR. Perfect security would require setting the FMR to 0, in which case the FNMR would be 1. At the other extreme, setting the FNMR to 0 would result in an FMR of 1.

Vendors often use equal error rate (EER), an additional metric derived from FMR and FNMR, to describe the accuracy of their biometric systems. EER refers to the point at which FMR equals FNMR. Setting a system's threshold at its EER will result in the probability that a person is falsely matched equaling the probability that a person is falsely not matched. However, this statistic tends to oversimplify the balance between FMR and FNMR, because in few real-world applications is the need for security identical to the need for convenience.

Figure 1: The General Relationship between FMR and FNMR

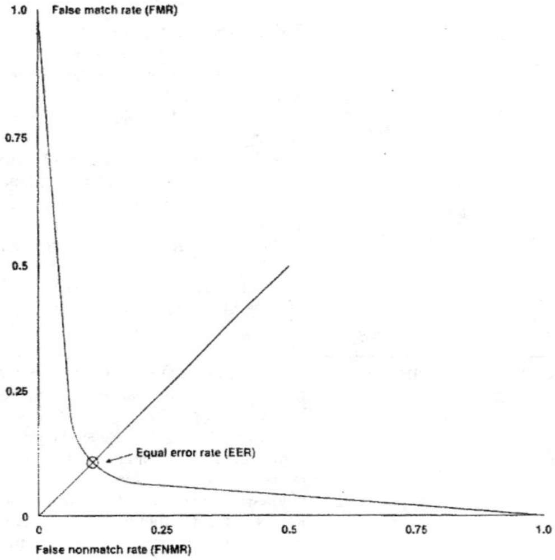

Source: GAO.

Note: Equal error rate is the point at which FMR equals FNMR.

FTER is a biometric system's third critical accuracy metric. FTER measures the probability that a person will be unable to enroll. Failure to enroll (FTE) may stem from an insufficiently distinctive biometric sample or from a system design that makes it difficult to provide consistent biometric data. The fingerprints of people who work extensively at manual labor are often too worn to be captured. A high percentage of people are unable to enroll in retina recognition systems because of the precision such systems require. People who are mute cannot use voice systems, and people lacking fingers or hands from congenital disease, surgery, or injury cannot use fingerprint or hand geometry systems. Although between 1 and 3 percent of the general public does not have the body part required for using any one biometric system, they are normally not counted in a system's FTER.

Using Multiple Biometrics

Because biometric systems based solely on a single biometric may not always meet performance requirements, the development of systems that integrate two or more biometrics is emerging as a trend. Multiple biometrics could be two types of biometrics, such as combining facial and iris recognition. Multiple biometrics could also involve multiple instances of a single biometric, such as 1, 2, or 10 fingerprints, 2 hands, and 2 eyes. One prototype system integrates fingerprint and facial recognition technologies to improve identification. A commercially available system combines face, lip movement, and speaker recognition to control access to physical structures and small office computer networks. Depending on the application, both systems can operate for either verification or identification. Experimental results have demonstrated that the identities established by systems that use more than one biometric could be more reliable, be applied to large target populations, and improve response time.

Standards for Biometric Technology

Identifying, exchanging, and integrating information from different and perhaps unfamiliar sources and functions are essential to an effective biometrics application. Without standards, system developers may need to define in detail the precise steps for exchanging information, a potentially complex, time-consuming, and expensive process. Progress has been made in developing biometrics standards. However, the majority of biometric devices and their software are still proprietary in many respects. For example, the method for extracting features from a biometric sample, such as a fingerprint, differs among most, if not all, vendors. Devices from company A do not necessarily work compatibly with devices from companies B and C.

Standards such as the National Institute of Science and Technology's (NIST) Common Biometric Exchange File Format (CBEFF) facilitate data exchange between different system components and simplify the integration of software and hardware from different vendors. The wavelet scalar quantization (WSQ) gray-scale fingerprint image compression algorithm is the standard for exchanging fingerprint images within the criminal justice system. Similarly, the Joint Photographic Experts Group (JPEG) has established an image compression standard that is designed to facilitate the transfer of images for facial recognition systems.

The American Association for Motor Vehicle Administration (AAMVA) included a format for fingerprint minutiae data in its Driver License and

Identification Standard, which provides a uniform means to identify issuers and holders of driver's licenses in the United States and Canada. However, the standard still allows for including data in a vendor-specific format. Biometric templates, which capture only the critical data needed to make a match, are small, but the template one vendor uses cannot generally be used by another for some biometric technologies, such as fingerprints. Without the creation and industry adoption of a biometric template standard, it could be necessary to store the larger biometric sample as well as the biometric template for each user during enrollment. Last year, the International Civil Aviation Organization (ICAO) New Technologies Working Group concluded that the only reliable globally interoperable method for exchanging face, fingerprint, or iris biometric data was the storage of the respective image. ICAO is studying the use of biometrics in machine-readable travel documents, such as passports and visas.

In November 2001, the executive board of the International Committee for Information Technology Standards (INCITS) established a technical committee for biometrics for the rapid development and approval of formal national and international generic biometric standards. Four task groups were created to conduct the work. The first task group is focused on the standardization of the content, meaning, and representation of biometric data interchange formats. This task group is working on formats for representing fingerprints, faces, irises, hand geometry, and signatures. The second task group covers the standardization of interfaces and interactions between biometric components and subsystems. CBEFF is an example of an interface standard. The third task group focuses on the development of biometric application profiles. It currently has projects in the areas of border crossings, transportation workers, and point of sale. The fourth task group handles the standardization of biometric performance metric definitions and calculations, approaches to test performance, and requirements for reporting the results of these tests.

Using Biometrics for Aviation Security

The Federal Aviation Administration (FAA), and subsequently, DHS and TSA, has been examining the use of biometrics for aviation security for several years. In 2001, the FAA and the Department of Defense Counterdrug Technology Development Program Office co-chaired the Aviation Security Biometrics Working Group (ASBWG). They examined the use of biometrics in four aviation security applications: (1) identity verification of employees and ensuring that access to secured areas within an airport is restricted to authorized personnel; (2) protection of public areas in and around airports using surveillance; (3) identity verification of passengers boarding aircraft; and (4) identity verification of flight crews prior to and during a flight. Subsequently, in 2002, TSA contracted with the International Biometric Group to evaluate the use of biometrics for automated surveillance within airports, trusted traveler cards for passengers, and identity verification of employees for access control in airports.[2]

Since the 2001 terrorist attacks, the Congress has directed a greater use of biometrics. For example, the Aviation and Transportation Security Act (ATSA), which created TSA and mandated several actions designed to enhance aviation security, includes several provisions regarding the use of biometrics for applications, such as perimeter security or access control.[3]

Access Control

Biometric systems have long been used to complement or replace badges and keys in controlling access to entire facilities or specific areas within a facility. The entrances to more than half the nuclear power plants in the United States employ hand geometry systems. Further, recent reductions in the price of biometric hardware have spurred logical access control applications. Fingerprint, iris, and speaker recognition are replacing passwords to authenticate individuals accessing computers and networks. The Office of Legislative Counsel of the U.S. House of Representatives, for example, is using an iris recognition system to protect confidential files and working documents. Other federal agencies, including the Department of Defense, Department of Energy, and Department of Justice, as well as the intelligence community, are adopting similar technologies.

We have previously reported on the critical need to limit access to secure airport areas. In 2000, we reported on the ability of our special agents to use fictitious law enforcement badges and credentials to gain access to secure areas of two commercial airports.[4] The agents, who had been issued tickets and boarding passes, were not screened through magnetometers at the security checkpoints nor was their baggage inspected. This vulnerability could have allowed our agents to carry weapons, explosives, or other dangerous objects onto an aircraft.

[2]International Biometric Group, "Framework for Evaluating and Deploying Biometrics in Air Travel Applications: Surveillance, Trusted Travel, Access Control" (Apr. 3, 2002).

[3]Aviation and Transportation Security Act (Public Law 107-71, Nov. 19, 2001).

[4]U.S. General Accounting Office, *Security: Breaches at Federal Agencies and Airports*, GAO/T-OSI-00-10 (Washington, D.C.: May 25, 2000).

Since 1991, San Francisco International Airport has used hand geometry devices in conjunction with identification cards to protect secure areas of the airport, such as the tarmac and loading gates. Last year, Toledo (Ohio) Express Airport also installed hand geometry devices to ensure that only authorized personnel can gain access to critical areas of the airport.

FAA has conducted several tests and pilots of biometrics for access control to secure areas of airports. In 1998, FAA funded an operational test at Chicago's O'Hare International Airport involving smart cards and fingerprint recognition to identify employees of motor carrier and air cargo companies at access control points to cargo areas. Further, in 2001, FAA conducted tests of hand geometry and fingerprint and facial recognition technologies for employee access control at airports.

TSA has two current efforts examining the use of biometrics for access control. The Transportation Worker Identification Credential (TWIC) is designed to be a common credential for all transportation workers requiring unescorted physical access to secure areas of the national transportation system, such as airports, seaports, and railroad terminals. It will also be used to help secure logical access to computers, networks, and applications. The program was developed in response to ATSA and the Maritime Transportation Security Act of 2002 and will include the use of biometrics to provide a positive match of a credential for up to 6 million transportation workers across the United States.[5] The TWIC program is designed as an identity authentication tool for individual facilities and to provide assurance that individuals with a TWIC card have undergone a threat assessment to ensure that they are not known terrorists. Individual facilities will be able to use the TWIC cards to control access to secure areas to only authorized individuals.

Last week, TSA issued a request for proposal for a TWIC prototype to determine the performance of TWIC as an access control tool. For the prototype, TSA will be examining the use of at least fingerprint and iris recognition. During a technology evaluation last year, TSA evaluated six card technologies and determined that an integrated circuit chip smart card was the most appropriate for the TWIC card. As part of the prototype, TSA will also examine the use of cards with 2-dimensional bar codes and optical stripes. The prototype phase is expected to last 7 months and will

[5] Aviation and Transportation Security Act, §106(c) and §136, and Maritime Transportation Security Act of 2002 (Public Law 107-295, Nov. 25, 2002), §102.

be conducted in Philadelphia, PA; Wilmington, DE; the ports of Long Beach and Los Angeles, CA; and the 14 major port facilities in the state of Florida. TSA anticipates that up to 200,000 workers will be enrolled in the program. Following the prototype, TSA will make a decision on whether to proceed with implementation of the program.

Earlier this month, TSA announced an access control pilot program that will test various technologies, including biometrics, that are designed to ensure that only authorized personnel have access to non-passenger controlled areas. Developed in response to a section in ATSA that directed the establishment of pilot programs to test and evaluate technologies for providing access control to closed or secure areas of airports, the program will test fingerprint recognition at four airports and iris recognition at one airport.[6] Boise Air Terminal/Gowen Field Airport, Southwest Florida International Airport, and Tampa International Airport will test fingerprint recognition to control vehicle access. Newark International Airport will test fingerprint recognition to allow only authorized persons into secure areas of the airport. T.F. Green State Airport (Providence, RI) will test iris recognition to control access to secure areas of the airport.

Registered Traveler

The concept of a registered traveler program is to provide an expedited security screening for passengers who meet the eligibility criteria and who voluntarily provide personal information and clear a background check. ATSA permits TSA to "establish requirements to implement trusted passenger programs and use available technologies to expedite the security screening of passengers who participate in such programs, thereby allowing security screening personnel to focus on those passengers who should be subject to more extensive screening."[7]

In 2002, we reviewed the policy and implementation issues associated with a registered traveler program.[8] We identified four key questions that need to be addressed by the federal government before proceeding with such a program: (1) What criteria should be established to determine eligibility to apply for the program? (2) What kinds of background checks should be

[6]Aviation and Transportation Security Act, §106(d).

[7]Aviation and Transportation Security Act, §109(a)(3).

[8]U.S. General Accounting Office, *Aviation Security: Registered Traveler Program Policy and Implementation Issues*, GAO-03-253 (Washington D.C.: Nov. 22, 2002).

used to certify that applicants are eligible to enroll in the program, and who should perform these? (3) Which security-screening procedures should registered travelers undergo, and how should these differ from those used for unregistered travelers? and (4) To what extent do equity, privacy, and liability issues have to be resolved prior to program implementation?

In April 2004, TSA issued a combined solicitation synopsis for a registered traveler pilot program. TSA has evaluated the capabilities statements from about 40 proposals. TSA expects to award contracts for the pilot program in early June 2004. The pilot program will run for about 90 days at up to five airports. TSA expects to enroll up to 10,000 travelers in the program using fingerprint and/or iris recognition. To enroll, travelers will submit biographic and biometric data at the selected airports. A security assessment will be conducted on the applicants to verify their eligibility for the program. TSA may use a TSA-issued card or an airline frequent flier card as an identifier to conduct biometric verification matches of registered travelers at airport security checkpoints. TSA is also considering the use of identification (1-to-many) matching to ascertain the identity of the registered traveler. Once registered travelers are identified, they will undergo an adjusted screening process, designed to expedite throughput for low-risk travelers.

Similar programs have been used for expediting border control processes. For example, the Immigration and Naturalization Service (INS) Passenger Accelerated Service System (INSPASS), a pilot program in place since 1993, has more than 45,000 frequent fliers enrolled at nine airports, and has admitted more than 300,000 travelers. It is open to citizens of the United States, Canada, Bermuda, and visa waiver program countries who travel to the United States on business three or more times a year.[9] To participate, users provide a passport or travel document and submit two fingerprints and a hand geometry biometric. Once travelers successfully undergo a background screening and are enrolled, they can circumvent immigration procedures and lines. An INSPASS participant presents their hand geometry biometric at an airport kiosk for comparison against the reference template stored in a central database for that traveler. INSPASS has reduced the inspection time for participants to less than 15 seconds.

[9]The visa waiver program permits nationals from designated countries to apply for admission to the United States for 90 days or less as nonimmigrant visitors for business or pleasure without first obtaining a U.S. nonimmigrant visa.

Airport Surveillance

It has been suggested that facial recognition could be used in airports as a surveillance tool that could identify persons of interest without the subject's cooperation or knowledge. Key to such an effort is the availability of a database of biometric information of persons of interest (i.e., a watch list). Surveillance activities are often conducted by humans who are looking for persons of interest using closed-circuit televisions. However, because it is well understood that humans are limited in their ability to recognize individuals they are not familiar with, and that there are limits of human attention when conducting surveillance activities, facial recognition has been cited as a potential surveillance tool.

In 2001, the ASBWG found that facial recognition technology was not sufficiently mature to be relied upon for wide-area surveillance. Further, as we reported in 2002, one vendor conducted pilots using facial recognition technology to conduct surveillance at U.S. airports. For these pilots, video cameras were installed at the security checkpoints, near the magnetometers. From the pilots, it was learned that lighting was the primary factor in determining the performance of facial recognition.

Other Federal Biometric Applications

There are two other primary uses of biometrics in the federal government: criminal identification and border security.

Criminal Identification

Fingerprint identification has been used in law enforcement over the past 100 years and has become the de facto international standard for positively identifying individuals. The Federal Bureau of Investigation (FBI) has been using fingerprint identification since 1928. The first fingerprint recognition systems were used in law enforcement about 4 decades ago.

The FBI's Integrated Automated Fingerprint Identification System (IAFIS) is an automated 10-fingerprint matching system that stores rolled fingerprints. The more than 40 million records in its criminal master file are connected electronically with all 50 states and some federal agencies. IAFIS was designed to handle a large volume of fingerprint checks against a large database of fingerprints. In 2002, we found that IAFIS processes, on average, approximately 48,000 fingerprints per day and has processed as many as 82,000 in a single day. IAFIS's target response time for criminal fingerprints submitted electronically is 2 hours; for civilian fingerprint background checks, 24 hours.

Border Security

There are several uses of biometrics for border security in the United States and worldwide.[10] Two notable examples are the INS Automated Biometric Fingerprint Identification System (IDENT) and the United States Visitor and Immigrant Status Indicator Technology (US-VISIT) system.

INS began developing IDENT around 1990 to identify illegal aliens who are repeatedly apprehended trying to enter the United States illegally. INS's goal was to enroll virtually all apprehended aliens. IDENT can also identify aliens who have outstanding warrants or who have been deported. When such aliens are apprehended, a photograph and two index fingerprints are captured electronically and queried against three databases. In 2002, IDENT had over 4.5 million entries. A fingerprint query of IDENT normally takes about 2 minutes.

Laws passed since the September 11, 2001, terrorist attacks require a more extensive use of biometrics for border control.[11] The Attorney General and the Secretary of State jointly, through NIST are to develop a technology standard, including biometric identifier standards.[12] When developed, this standard is to be used to verify the identity of persons applying for a U.S. visa for the purpose of conducting a background check, confirming identity, and ensuring that a person has not received a visa under a different name. Further, aliens are to be issued machine-readable, tamper-resistant visas and other travel and entry documents that use biometric identifiers. Similarly, equipment and software are to be installed at all ports of entry that can allow the biometric comparison and authentication of all U.S. visas and other travel and entry documents issued to aliens and machine-readable passports.

[10]We describe several of these uses in *Technology Assessment: Biometrics for Border Security*, GAO-03-174.

[11]See the Uniting and Strengthening America by Providing Appropriate Tools Required to Intercept and Obstruct Terrorism Act of 2001 (USA PATRIOT Act) (Public Law 107-56, Oct. 26, 2001), §403(c) and §414, and the Enhanced Border Security and Visa Entry Reform Act of 2002 (Public Law 107-173, May 14, 2002), §202(a)(4) and §303.

[12]In January 2003, in response to this requirement, NIST submitted its technical standards for biometric identifiers and tamper-resistance for travel documents as a part of a joint report to the Congress from the Attorney General, the Secretary of State, and NIST. NIST recommended that 10 fingerprints be used for background identification checks and that a dual biometric system using 2 fingerprint images and a face image may be needed to meet projected system requirements for verification.

DHS is developing the US-VISIT system to address these requirements. The US-VISIT system currently uses IDENT technology to collect a photograph and two index fingerprints from travelers holding non-immigrant visas. Travelers are initially enrolled either at a port of entry using US-VISIT entry procedures or at a U.S. consulate or embassy when they apply for their visa. US-VISIT entry procedures are currently in place at 115 airports and 14 seaports. By December 31, 2004, US-VISIT is planned to be in place at the 50 busiest land ports of entry. By December 31, 2005, US-VISIT is planned to be in place at all 165 land ports of entry. As of March 4, 2004, biometric data collection was in place at more than 80 visa-adjudicating posts. By October 2004, biometric data collection is expected to be in use at all 211 visa-issuing embassies and consulates. By September 30, 2004, US-VISIT procedures will be expanded to include visitors traveling to the United States under the visa waiver program arriving at air and sea ports of entry.

Each time a visitor enters the United States at a port of entry employing US-VISIT entry procedures, the visitor's fingerprints will be matched against the reference fingerprints captured during enrollment. During enrollment and each subsequent visit, the biographic and biometric data of the visitor is compared to watch lists to assist the inspectors in making admissibility decisions. At one airport and one seaport, visitors are also expected to record their departure from the United States using an automated self-service kiosk that can scan the visitor's travel documents and capture the visitor's fingerprints.[13]

Challenges and Issues in Using Biometrics

While biometric technology is currently available and used in a variety of applications, questions remain regarding the technical and operational effectiveness of biometric technologies in large-scale applications. We have found that a risk management approach can help define the need and use for biometrics for security. In addition, a decision to use biometrics should consider the costs and benefits of such a system and its potential effect on convenience and privacy.

[13]GAO has conducted reviews of annual expenditure plans of the US-VISIT program. The review of the fiscal year 2004 expenditure plan can be found in U.S. General Accounting Office, *Homeland Security: First Phase of Visitor and Immigration Status Program Operating, but Improvements Needed*, GAO-04-586 (Washington, D.C.: May 11, 2004).

Risk Management Is the Foundation of Effective Strategy

The approach to good security is fundamentally similar regardless of the assets being protected. As we have previously reported, these principles can be reduced to five basic steps that help to determine responses to five essential questions (see figure 2).[14]

Figure 2: Five Steps in the Risk Management Process

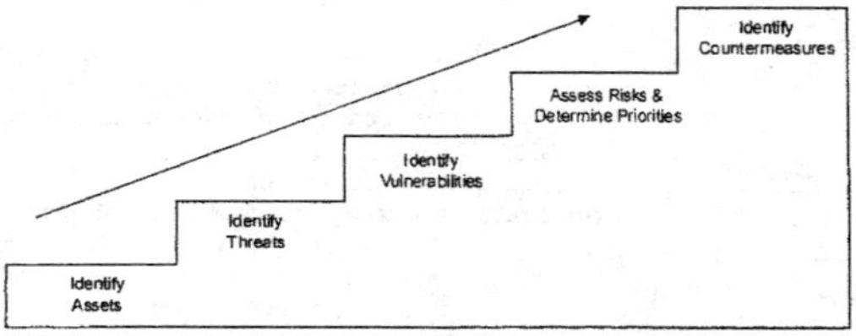

Source: GAO.

What Am I Protecting?

The first step in risk management is to identify assets that must be protected and the impact of their potential loss.

Who Are My Adversaries?

The second step is to identify and characterize the threat to these assets. The intent and capability of an adversary are the principal criteria for establishing the degree of threat to these assets.

How Am I Vulnerable?

The third step involves identifying and characterizing vulnerabilities that would allow identified threats to be realized. In other words, what weaknesses can allow a security breach?

[14]U.S. General Accounting Office, *National Preparedness: Technologies to Secure Federal Buildings*, GAO-02-687T (Washington, D.C.: Apr. 25, 2002).

What Are My Priorities?

In the fourth step, risk must be assessed and priorities determined for protecting assets. Risk assessment examines the potential for the loss or damage to an asset. Risk levels are established by assessing the impact of the loss or damage, threats to the asset, and vulnerabilities.

What Can I Do?

The final step is to identify countermeasures to reduce or eliminate risks. In doing so, the advantages and benefits of these countermeasures must also be weighed against their disadvantages and costs.

Protection, Detection, and Reaction Are Integral Security Concepts

Countermeasures identified through the risk management process support the three integral concepts of a holistic security program: protection, detection, and reaction. Protection provides countermeasures such as policies, procedures, and technical controls to defend against attacks on the assets being protected. Detection monitors for potential breakdowns in protective mechanisms that could result in security breaches. Reaction, which requires human involvement, responds to detected breaches to thwart attacks before damage can be done. Because absolute protection is impossible to achieve, a security program that does not incorporate detection and reaction is incomplete.

Biometrics can support the protection component of a security program. It is important to realize that deploying them will not automatically eliminate all security risks. Technology is not a solution in isolation. Effective security also entails having a well-trained staff to follow and enforce policies and procedures. Weaknesses in the security process or failures by people to operate the technology or implement the security process can diminish the effectiveness of technology.

Accordingly, there is a need for the security process to account for limitations in technology. For example, procedures for exception processing would also need to be carefully planned. As we described, not all people can enroll in a biometrics system. Similarly, false matches and false nonmatches will also sometimes occur. Procedures need to be developed to handle these situations. Exception processing that is not as good as biometric-based primary processing could be exploited as a security hole. The effect on the process is directly related to the performance of the technology. In our study of biometrics for border security, we found that fingerprint recognition appears to be the most mature of the biometric technologies. Fingerprint recognition has been used the longest and has been used with databases containing up to 40 million entries. Iris recognition is a young technology and has not been used with large populations. While facial recognition has also been used with large databases, its accuracy results in testing have lagged behind those of iris and fingerprint recognition.

As with any credentialing or identity management system, it is critical to consider the process used to issue the credential. Biometrics can help ensure that people can only enroll into a security system once and to ensure that a person presenting himself before the security system is the same person that enrolled into the system. However, biometrics cannot necessarily link a person to his or her true identity. While biometrics would make it more difficult for people to establish multiple identities, if the one identity a person claimed were not his or her true identity, then the person would be linked to the false identity in the biometric system. The use of biometrics does not relieve the credential-issuing authority of the responsibility of ensuring the identity of the person requesting the credential or of conducting a security check, commensurate with the level of access being granted, to assure itself that the person is entitled to receive the credential. The quality of the identifier presented during the enrollment process is key to the integrity of a biometrics system.

Even if the biometric is checked against a biometrics-based watch list, the effectiveness of such a list is also dependent on nontechnological processes. The policies and procedures governing the population of the watch list as well as the effectiveness of the law enforcement and intelligence communities to identify individuals to place on the watch list are critical to the success of the program. People who are not on the watch list cannot be flagged as someone who is not eligible to receive a credential.

Deciding to Use Biometric Technology

A decision to use biometrics in a security solution should also consider the benefits and costs of the system and the potential effects on convenience and privacy.

Weighing Costs and Benefits

Best practices for information technology investment dictate that prior to making any significant project investment, the benefit and cost information of the system should be analyzed and assessed in detail. A business case should be developed that identifies the organizational needs for the project and a clear statement of high-level system goals should be developed. The high-level goals should address the system's expected outcomes such as the binding of a biometric feature to an identity or the identification of undesirable individuals on a watch list. Certain performance parameters should also be specified such as the time required to verify a person's identity or the maximum population that the system must handle.

Once the system parameters are developed, a cost estimate can be developed. Not only must the costs of the technology be considered, but also the costs of the effects on people and processes. Both initial costs and recurring costs need to be estimated. Initial costs need to account for the engineering efforts to design, develop, test, and implement the system; training of personnel; hardware and software costs; network infrastructure improvements; and additional facilities required to enroll people into the biometric system. Recurring cost elements include program management costs, hardware and software maintenance, hardware replacement costs, training of personnel, additional personnel to enroll or verify the identities of people in the biometric system, and possibly the issuance of token cards for the storage of biometrics.

Weighed against these costs are the security benefits that accrue from the system. Analyzing this cost-benefit trade-off is crucial when choosing specific biometrics-based solutions. The consequences of performance issues—for example, accuracy problems, and their effect on processes and people—are also important in selecting a biometrics solution.

Effects on Privacy and Convenience

The Privacy Act of 1974 limits federal agencies' collection, use, and disclosure of personal information, such as fingerprints and photographs.[15] Accordingly, the Privacy Act generally covers federal agency use of personal biometric information. However, the act includes exemptions for law enforcement and national security purposes. Representatives of civil liberties groups and privacy experts have expressed concerns regarding (1) the adequacy of protections for security, data sharing, identity theft, and other identified uses of biometric data and (2) secondary uses and "function creep." These concerns relate to the adequacy of protections under current law for large-scale data handling in a biometric system. Besides information security, concern was voiced about an absence of clear criteria for governing data sharing. The broad exemptions of the Privacy Act, for example, provide no guidance on the extent of the appropriate uses law enforcement may make of biometric information.

[15] 5 U.S.C. §552a.

Because there is no general agreement on the appropriate balance of security and privacy to build into a system using biometrics, further policy decisions are required. The range of unresolved policy issues suggests that questions surrounding the use of biometric technology center as much on management policies as on technical issues.

Finally, consideration must be given to the convenience and ease of using biometrics and their effect on the ability of the agency to complete its mission. For example, some people find biometric technologies difficult, if not impossible, to use. Still others resist biometrics because they believe them to be intrusive, inherently offensive, or just uncomfortable to use. Lack of cooperation or even resistance to using biometrics can affect a system's performance and widespread adoption.

Furthermore, if the processes to use biometrics are lengthy or erroneous, they could negatively affect the ability of the assets being protected to operate and fulfill its mission. For example, in 2002, we found that there are significant challenges in using biometrics for border security. The use of biometric technologies could potentially impact the length of the inspection process. Any lengthening in the process of obtaining travel documents or entering the United States could affect travelers significantly. Delays inconvenience travelers and could result in fewer visits to the United States or lost business to the nation. Further studies could help determine whether the increased security from biometrics could result in fewer visits to the United States or lost business to the nation, potentially adversely affecting the American economy and, in particular, the border communities. These communities depend on trade with Canada and Mexico, which totaled $653 billion in 2000.

In conclusion, biometric technologies are available today that can be used for aviation security. However, it is important to bear in mind that effective security cannot be achieved by relying on technology alone. Technology and people must work together as part of an overall security process. As we have pointed out, weaknesses in any of these areas diminishes the effectiveness of the security process. We have found that three key considerations need to be addressed before a decision is made to design, develop, and implement biometrics into a security system:

1. Decisions must be made on how the technology will be used.

2. A detailed cost-benefit analysis must be conducted to determine that the benefits gained from a system outweigh the costs.

3. A trade-off analysis must be conducted between the increased security, which the use of biometrics would provide, and the effect on areas such as privacy and convenience.

Security concerns need to be balanced with practical cost and operational considerations as well as political and economic interests. A risk management approach can help federal agencies identify and address security concerns. To develop security systems with biometrics, the high-level goals of these systems need to be defined, and the concept of operations that will embody the people, process, and technologies required to achieve these goals needs to be developed. With these answers, the proper role of biometric technologies in aviation security can be determined. If these details are not resolved, the estimated cost and performance of the resulting system will be at risk.

Mr. Chairman, this concludes my statement. I would be pleased to answer any questions that you or members of the subcommittee may have.

Contacts

For further information, please contact Keith Rhodes at (202)-512-6412 or Richard Hung at (202)-512-8073.

TESTIMONY OF C. STEWART VERDERY, JR
ASSISTANT SECRETARY FOR BORDER AND TRANSPORTATION SECURITY
POLICY AND PLANNING
DEPARTMENT OF HOMELAND SECURITY
BEFORE THE HOUSE AVIATION SUBCOMMITTEE
May 19, 2004

Chairman Mica and other distinguished Members, it is a pleasure to appear before you today to discuss how the Department of Homeland Security (DHS) is using biometrics to improve aviation security and facilitate legitimate trade and travel, particularly in the US-VISIT and Transportation Worker Identity Credential (TWIC) programs.

No single problem area has mobilized government and private sector activity in the area of identification systems as much as global terrorism. More than two years after September 11, exploration and actual implementation of the use of biometrics to ensure identity and enhance security continues to be an area of fevered activity in both the domestic and international arenas.

Biometrics is the science of identifying, recording and matching unique physical characteristics to individuals. There are five basic technologies: facial recognition, fingerprint, hand geometry, iris recognition and voice recognition.

The creation of DHS has allowed agencies to rethink security procedures and, in many cases, adapt IT infrastructures to include new biometric technologies.

US-VISIT is a Border and Transportation Security (BTS) program that represents a continuum of security measures that uses biometrics as a key element. Both State and DHS use biometrics and biographic data to check individual visa applicants against appropriate "lookout" data. In addition, these biometric technologies such as digital, inkless fingerscans and digital photographs also enable DHS to determine whether the person applying for entry to the United States is the same person who was issued a visa by the State Department.

Areas where DHS is currently exploring the potential for biometrics to enhance aviation security include Transportation Security Administration's (TSA) recently announced award of grants to explore using biometric technologies to enhance airport security access controls, TSA's Registered Traveler Pilot Program, and the Transportation Worker Identification Credential program. For all of these programs, TSA is working with TSA's privacy officer and the DHS privacy officer to ensure that all relevant privacy considerations are taken into account.

TSA's will test Anti-Piggybacking technology (technology that would prevent someone from gaining access through a control point by following immediately after with someone else's identification) and other technologies, advanced video surveillance technology and various biometric technologies to ensure that only authorized personnel have access to non-passenger controlled areas. Under the TWIC program, TSA is assessing how smart card technology that incorporates a biometric feature could be used to enhance the security of transportation facilities nationwide, including seaports, rail and transit facilities, airports and others. TSA's Registered Traveler Pilot program is enabling TSA to explore technological solutions associated with positive identity verification, including biometrics, to facilitate the movement of passengers who have received a prior security assessment through airport security checkpoints.

I will discuss all of these initiatives in greater detail below.

US-VISIT

In 1996 and 2000, the United States Congress mandated the creation of an electronic entry-exit system to manage the entry and departure of foreign visitors. After the events of September 11, 2001, Congress added the requirement that the entry exit system have the capability to confirm identity. DHS has established the US-VISIT Program to accomplish these statutory mandates and to achieve the following goals:

- Enhance the safety of our citizens and visitors;
- Facilitate legitimate travel and trade;
- Ensure the integrity of our immigration system; and
- Protect the privacy of travelers to the United States.

US-VISIT represents a major milestone in enhancing our nation's security and our efforts to reform our borders. It is a significant step towards bringing integrity back to our immigration and border enforcement systems. It is also leading the way for incorporating biometrics into international travel security systems.

US-VISIT is a continuum of security measures that begins before individuals enter the United States and continues through their arrival and departure from the country. Enrolling travelers in US-VISIT using biometric identifiers allows DHS to:

- Conduct appropriate security checks: We conduct checks of visitors against appropriate lookout databases available to consular officers and inspectors, including biometric-based checks.
- Freeze identity of traveler: We biometrically enroll visitors in US-VISIT – freezing the identity of the traveler and tying that identity to the travel document presented.
- Match traveler identity and document: We biometrically match that identity and document, with the information collected by State, enabling the inspector to determine whether the traveler complied with the terms of her/his previous admission and is using the same identity.
- Document arrivial, and departure: We collect automated arrival and departure information on travelers.
- Determine overstays: We will use collected information to determine whether individuals have overstayed the terms of their admission. This information will be

used to determine whether an individual should be apprehended or whether the individual should be allowed to enter the U.S. upon her/his next visit.

The Department of Homeland Security with the Department of State Consular Affairs have created an entire continuum of identity verification measures that begins overseas collecting fingerprints, when a traveler applies for a visa, and continues upon entry and exit from this country. The system stores biometric and biographic data in a secure, centralized database and uses travel and identity documents to access that information for identity verification and watchlist checks. Today, more than 130 visa-issuing posts have begun to capture fingerscans and digital photographs of foreign nationals when they apply for visas, regardless of their country of origin. We expect that this process will be in place at all 211 visa-issuing posts worldwide by October 2004.

At the U.S. border, certain visitors are required to provide biometric data, biographic data, and/or other documentation. This data is checked against US-VISIT databases, which contain visa issuance information, watchlists, including information from the Federal Bureau of Investigation, and immigration status information allowing border inspectors to verify identity and identify security threats and immigration violators. In its first 4 months of operation, DHS processed nearly 3.65 million foreign national applicants for admission through US-VISIT at its air and sea ports of entry. During that period, 291 individuals were identified by biometrics alone as being the subject of a lookout. DHS took adverse action in 43% of the 291 cases. Of the 291 hits, 62% were for criminal violations (some of which were immigration related criminal violations, such as previous deportation); 38% were for immigration violations alone.

One of the US-VISIT Program's primary roles is to identify those individuals who have overstayed the terms of their admission. Currently, our exit procedures are based upon receiving departure information from passenger manifests shared with us by air carriers under Custom and Border Protection's (CBP) Advanced Passenger Information System (APIS). We match information received under APIS with admission information when a passenger applies for entry into the U.S. at the port of entry, and identify those likely to have overstayed the terms of their admission. We are testing our ability to enhance matching of arrival and departure records by using biometrics in various pilot programs, one of them being at the Baltimore-Washington International Airport. We plan to expand our pilot program to a total of 15 air and seaports over the next several months. We will pilot test three options and evaluate the results to identify the best, most efficient and effective process.

At various points in the pre-entry, entry, status management and exit processes, decision makers are supported by systems checks against data from law enforcement and intelligence sources that identify persons of interest for various violations. All names and fingerscans are checked against watch lists to identify known or suspected terrorists, criminals and immigration violators.

In just a few months, the first release of US-VISIT has improved the security of our citizens and visitors. Our CBP Officers are saying that the new tools we have put in place truly help them do their jobs more effectively and are a major advancement in border control. US-VISIT adds, on average, only fifteen seconds to the average inspection time. Included in this processing time are the collection of the biometric and biographic information, the comparison of that information with that collected by the Department of State at the time of visa issuance, and the screening of the biographic and biometric information through watchlists and other criminal history information.

US-VISIT is working. We intercepted a fugitive who had escaped from prison over 20 years ago. We caught and extradited a felon wanted for manslaughter in San Diego. We finally stopped one drug dealer who had entered the U.S. more than 60 times in the past four years using different names and dates of birth. We continue to identify criminals every day at our borders, and since January 19, we have supplied crucial biometric information to our partners at the Department of State to help prevent ineligible visa applicants from obtaining a visa.

The increase in security has not had negative effect on our wait times or our commitment to service. But you don't have to take my word for it. Albert Park, a Korean visiting his sister and arriving at John F. Kennedy International Airport, told the New York Sun (January 6th edition): "I expected a lot more delays, but it was all pretty smooth." He went on to state that "[US-VISIT] definitely makes me feel safer."

"We at the airport believe that this is a true enhancement," said Bruce Drum, associate director of the Miami-Dade County Aviation Department." (The Associated Press, January 5th)

The Washington Post (January 6th) also reported on travelers' perceptions of the additional security measures: "Some travelers who were fingerprinted and photographed at airports across the country yesterday said the security procedures were swift, and most said they were resigned to the new rules. 'I don't really mind,' said D.C. resident Salome Nnanga, a native of Ethiopia. 'I think it's a very, very good idea to protect the country.'"

We want to ensure that we continue to be a welcoming nation, a nation that invites visitors to study, do business, and relax in our country. We also owe it to our citizens and visitors to deny entry to persons wishing to do harm, or who are inadmissible to the U.S. Few would dispute that these steps are necessary.

As we evaluate the first four months of the program, it seems clear that visitors appreciate the effort we are making to deliver security while simultaneously facilitating the process for law-abiding, legitimate travelers. We must continue to respect our visitors' privacy, treat them fairly, and enable them to pass through inspection quickly so they can enjoy their visit in our country. As people attempt to enter our country, we must know who they are and whether we have information that they have committed a crime that would make them inadmissible to the U.S. The ability of US-VISIT to rapidly screen applicants, using biometrics, means we can have security and control without impeding legitimate travelers, and we can also help protect our welcomed visitors by drastically reducing the possibility of identity theft. Moreover, as visitors leave the country, we must know that they have not overstayed their period of authorized stay.

But we are not finished. This is a complicated job that will take time to complete. In fact, US-VISIT is designed to be rolled out in increments to ensure that the foundation is strong and the building blocks are effective. With the deployment of the entry components at air and sea ports, we have made a strong beginning, and we plan to meet the December 31, 2004, deadline to deploy US-VISIT at the 50 busiest land border ports of entry. We also expect to deploy biometric capabilities at those ports of entry to allow DHS to check the identity of certain travelers against watchlists and databases. We are seeing that we can accomplish what we set out to do: keep out criminals and terrorists, enhance the integrity of our immigration system, facilitate legitimate travel and trade and help protect the privacy and identity of our visitors.

An obvious concern for all legitimate travelers is that criminals may use their lost or stolen travel documents to enter the United States. Biometric identifiers make it difficult for criminals to travel on someone else's travel documents. This is a significant benefit that US-VISIT delivers for the millions of legitimate travelers we welcome each year. In addition, we must continue to respect our visitors' privacy. We have a Privacy Impact Assessment (PIA) being reviewed by external audiences and DHS has the first statutorily created Chief Privacy Officer, Nuala O'Connor Kelly. Ms. O'Connor Kelly along with the US-VISIT privacy officer has worked closely with privacy experts at the Office of Management and Budget, and with independent privacy consultants to prepare a PIA that addresses the beginning increments of this program.

The Department is not doing this alone. We are collaborating with other government agencies, most notably the Department of State, to implement US-VISIT and inform the traveling public. We are working closely with the air and sea travel industry regarding the requirements of the US-VISIT program, as well as speaking with constituencies along the land borders. We see our relationship with these groups as a partnership.

We are also partnering with private industry to develop the best technological solutions. In accordance with our published schedule, a Request for Proposals (RFP) was issued in November 2003. The RFP incorporates an acquisition strategy to ensure that the latest available technologies will be incorporated into US-VISIT. We expect to award the contract for this technology later this month.

An important part of the program is public education. Travelers are educated about the program before they arrive at the port of entry. We are engaged in a worldwide campaign to inform them. This campaign includes public service announcements, signage at ports of entry, explanatory cards on airplanes and cruise ships, news media coverage and on-board explanatory videos.

US-VISIT is critical to our national security as well as our economic security, and its introduction has been successful. We are committed to building a system that enhances the integrity of our immigration system by catching the few and expediting the many, and we recognize that the United States is leading the way in helping other countries around the world keep their borders secure and their doors open.

AIRPORT ACCESS CONTROL PILOT PROGRAM (AACPP)

The second BTS program that is exploring biometrics technology is the Airport Access Control Pilot Program within TSA. The Aviation and Transportation Security Act (ATSA) required the establishment of pilot programs at no fewer than 20 airports to test and evaluate new and emerging technology for providing access control and other security protections for closed or secure areas of the airports. ATSA also states that the technologies to be evaluated under the pilot programs may include, among others, biometric technologies. To meet this requirement, TSA has developed a two-phase pilot program, for which awards for Phase I were recently announced.

Phase I of the pilot includes testing of various off-the-shelf technologies, including biometric technologies including fingerprint, under a variety of real-world operational environments. Based on that analysis, TSA will then determine which technologies will be evaluated in the Phase II airports. The Phase I pilot programs will focus on identifying the operational payoffs achievable through increased use of biometric and other technologies.

In selecting airports for participation, TSA began contacting airports in October 2002 to gauge their level of interest in the program. TSA asked the 82 airports that expressed preliminary interest to complete a survey so TSA could determine how well each applicant airport fit the desired characteristics and evaluate the airport authority and management's willingness to cooperate in the pilot. Of the 55 that responded to the surveys by October 28, 2003, TSA conducted further analysis and site surveys to choose airports for participation in phase I of the program.

In selecting technologies for assessment under the pilot program, TSA issued a request for information in December, 2002. More than 350 individuals submitted technology candidates for consideration.

For Phase I, which is funded at $8,000,000, TSA announced on May 3, 2004, the selection of eight airports:

- *Boise Air Terminal/Gowen Field Airport* will test a system that combines fingerprint biometric and Radio Frequency Identification (RFID) technology to control vehicle access.
- *Miami International Airport* will test a new defense system that will incorporate intelligent video analysis and other technology to detect intruders at the perimeter.
- *Minneapolis-St. Paul International Airport* will demonstrate a detection system using intelligent video analysis to differentiate between persons who are authorized and not authorized access to secured areas of the airport.
- *Newark International Airport* will test a system using fingerprint biometric technology to allow only authorized persons in secure areas of the airport.

- *Savannah International Airport* will focus on intelligent video surveillance technology to allow only authorized personnel to operate a cargo elevator that provides access to secure areas of the airport.
- *Southwest Florida International Airport* will evaluate new RFID and wireless fingerprint biometric technology intended to enhance the level of security at a vehicle gate.
- *T. F. Green State Airport* (in Providence, RI) will focus on controlling access to a secure area via an iris biometric recognition system. In addition, the entrance will employ anti-piggy backing detection (stopping more than one vehicle from gaining entrance at a time).
- *Tampa International Airport* will test the viability of portable card readers and fingerprint recognition technology at a vehicle gate.

Two additional airports will be selected at a later date, for a total of 10 Phase I airports. Various technology will be tested during Phase I including combining fingerprint biometric and Radio Frequency Identification (FRID) technology to control vehicle access; incorporating intelligent video analysis and other technology to detect intruders and unauthorized access; and controlling access to a secure area via an iris biometric recognition system. Phase I projects will be completed by December, 2004.

After Phase I and Phase II (which will expand on the number of technologies tested in additional airport operating environments) are both completed, information gathered during these pilot projects will be made available to appropriate airport and aviation industry representatives so that they may make informed decisions when designing access control systems to meet their security and regulatory needs. TSA will also make the results of these pilot projects available to other program areas within DHS, as well as other government agencies that may have a need for designing systems that provide facility security and/or establish programs using the various technologies evaluated, including biometric technologies. TSA and US-VISIT have collaborated closely to leverage expertise within the programs.

TRANSPORTATION WORKER IDENTIFICATION CREDENTIAL (TWIC)

TSA, through its TWIC program, is testing alternatives for developing and/or implementing a secure credential that could be used to mitigate potential threats posed by workers in transportation industries with fraudulent identification. The TWIC program is intended to enhance security controls applicable to the variety of transportation personnel whose duties require unescorted access to secure areas.

TSA has proceeded with development of the TWIC program in four phases. The first and second phases—planning and technology evaluation—have been completed. The technology evaluation phase consisted of testing at transportation facilities in the Philadelphia/Delaware River Basin and Los Angeles / Long Beach pilot program sites. During this phase, cards utilizing various technologies were issued to transportation workers with access to the facilities, and card technology performance data was collected. This phase led to TSA's selection of the Integrated Circuit Chip (ICC) technology as most suitable in the tested operational environments. The ICC is based on the National Institute of Standards and Technology (NIST) Government Smart Card Specification, includes encryption, secret keys, and active defenses, and can house a securely embedded biometric.

Phase III—the prototype phase—will involve evaluation of a broad range of business processes pertaining to identity management. These processes include enrollment of the applicants including the collection of biometrics, verification of claimed identity, and relevant checks of background information. Operational testing and evaluation will be conducted to select the biometric(s) to be used for the reference biometrics on the credential. A Request for Proposals (RFP) to begin the Prototype Phase of the Transportation Worker Identification Credential program was issued on May 10, 2004. The Prototype Phase is scheduled to last approximately seven months, and will be followed by Phase IV – implementation.

The TWIC program is being designed to leverage existing local facility control systems to the maximum extent possible, and has the potential to improve both commerce and security by providing "one credential mobility" across a number of different facilities. Decisions on how to implement a credentialing system will follow an assessment by DHS of the various prototype efforts. Our assessment will look at the cost and benefit of different approaches; most importantly how these benefit security.

REGISTERED TRAVELER PROGRAM

Finally, I will turn to the Registered Traveler (RT) Pilot Program, on which I know this Subcommittee has expressed keen interest even before enactment of ATSA and creation of the TSA.

As I mentioned before, TSA's pilot testing for a Registered Traveler program is designed to determine the feasibility of providing expedited movement through airport security checkpoints for travelers who volunteer to provide enough information about themselves to receive a security assessment indicating that they do not pose a threat to aviation security. Volunteers who participate in the RT Pilot program will also be requested to submit personal data, possibly including biometrics that will be used to validate identity using relevant government databases. Participants in the program will still be required to submit to screening for weapons, explosives, and prohibited items at the checkpoint.

TSA has collaborated with key internal and external stakeholders regarding the feasibility of such a program. Based upon interest expressed, TSA intends to conduct RT Pilots at a limited number of airports beginning in June, 2004. The pilots will last approximately 90 days. On April 5, 2004, TSA issued the first of a two-part Request for Proposal (RFP) soliciting input from the private sector for implementing Registered Traveler Pilots, and on May 13, 2004, TSA issued the second of a two-part RFP to those vendors that submitted the most highly rated capability statement to the initial Registered Traveler RFP. Awards for the Pilot operations will be made in mid-June 2004.

TSA awaits the results of the Pilot program prior to determining the feasibility and effectiveness of a broader implementation, including what costs, if any, would be incurred by those passengers who wish to participate in a future phase of the voluntary program. Upon conclusion of the pilots, results will be analyzed to ascertain security and customer service benefits and to determine the best approach for proceeding.

Conclusion

The Department is working, with its partners, to bring our nation's immigration and transportation security system into the 21st century. Technology must be utilized to move toward achieving the President's goal of secure U.S. borders and open doors to legitimate trade and travel.

Biometrics identifiers in the form of photographs and fingerprints have long played a key role in securing transportation systems and facilities; however human matching is subject to high costs and slow performance. The advent of automated matching capability gives us the ability to improve performance and permit the deployment and use of new technologies in new ways to assist us in freezing or fixing identities of foreign nationals, improve document security, and deter illegal access. In order to maximize our return on investment, it is vital that federal agencies and associated industries, also responsible for security of infrastructure, work together to create compatible systems.

REPORT

OF THE

2002 SILICON VALLEY BLUE RIBBON TASK FORCE

ON

AVIATION SECURITY AND TECHNOLOGY

JUNE 17, 2002

CONVENED BY

U.S. CONGRESSMAN MIKE HONDA

AND

SAN JOSE MAYOR RON GONZALES

CHAIRED BY

JOHN W. THOMPSON

CEO & CHAIRMAN, SYMANTEC CORPORATION

CONTENTS

June 17, 2002

The Honorable Mike Honda
Member of Congress, 15th District
3500 Stevens Creek Blvd.
San José, CA 95117

The Honorable Ron Gonzales
Mayor, City of San Jose
801 N. First Street, Suite 600
San José, CA 95110

Dear Congressman Honda and Mayor Gonzales,

On February 4, 2002, you announced formation of Blue Ribbon Task Force consisting of 20 technology, security, business and aviation experts to "identify and evaluate technology-driven solutions to improve the security and efficiency of national and local aviation." You asked the Task Force to complete its mission and report its findings within 100 days of its organization. Today, we are pleased to provide you the results of our work.

This report represents the contributions of some of the finest minds and forward-thinking companies in Silicon Valley. It contains a series of recommended technologies that can be implemented to enhance security and safety in the airport workplace, workforce and infrastructure communications network without jeopardizing the civil rights and civil liberties of the flying public. Most important, the technology proposals put forth by the Task Force can be put into place *now*.

The Task Force worked hard to develop its recommendations. During its brief 100-day period of existence, the Task Force and/or its two committees on Technology Demonstration and Report Writing met on more than a dozen occasions. The Task Force also held a public hearing and organized a meeting of vendors representing various technologies.

We believe that our proposals represent an important step toward infusing 21st Century technology into 21st Century air travel security, safety and comfort. We look forward to working with you in the future on any next steps required to transform these recommendations into reality.

Sincerely,

John W. Thompson
CEO and Chairman, Symantec
Chair, Blue Ribbon Task Force

Beatriz V. Infante
Chairman, President & CEO, Aspect Communications
Chair, Technology Demonstration Committee

Michael E. Fox, Sr.
Chairman, M.E. Fox & Co.
Chair, Report Writing Committee

LIST OF PARTICIPANTS

HONORARY CO-CHAIRS

Mike Honda, Member, U.S. Congress
Ron Gonzales, Mayor, City of San José

WORKING CHAIR

John W. Thompson, CEO and Chairman, Symantec

Task Force Committee Chairs
Michael E. Fox, Sr., Chairman, M.E. Fox and Company, Proposal Development
Beatriz V. Infante, CEO, Aspect Communications, Technology Recommendations

Committee Members
Sam Araki, Chairman, Security Technology Ventures
Captain Dan B. Ashby, Chair, California Airline Pilots Association
Bill Coleman, Chairman, BEA Systems, Inc.
Bill Crowell, President and CEO, Cylink Corporation
Mariano-Florentino Cuéllar, Assistant Professor, Stanford Law School
Sandra England, Executive Vice President, Network Associates
Don Harris, Director of Special Projects, Southwest Airlines
Gen. Richard Hearney (Ret.), CEO, BENS
Bill Lansdowne, Chief, San José Police
Dr. Sergio Magistri, CEO, InVision Technologies, Inc.
Bob McCashin, CEO, Identix
Dr. Ko Nishimura, CEO, Solectron Corporation
Richard W. Palmer, Jr., Vice President, Cisco Systems
Krish Panu, CEO, @Road
Larry A. Wansley, Managing Director of Corporate Security, American Airlines
Tom Weidemeyer, COO, United Parcel Service
Peggy Weigle, CEO, Sanctum

ii

STAFF TO THE TASK FORCE

Office of Congressman Mike Honda
Meri Maben, District Director
Matt Bostick, Field Representative

Office of Mayor Ron Gonzales
Jim Webb, Senior Policy Advisor
David Vossbrink, Communications Director

Office of City Manager, San José
Darrell Dearborn, Senior Deputy City Manager

Norman Y. Mineta San José International Airport
Ralph G. Tonseth, Director of Aviation
Jim Peterson, Deputy Director

Consultants
Larry N. Gerston, Ph.D., Gerston & Associates, Principal Writer
Sunny Claggett, Shilling & Kenyon, Task Force Facilitator
Callie Gregory Grant, Project Coordination

TASK FORCE RECOMMENDATIONS

NEW TECHNOLOGIES CAN IMPROVE THE NATION'S AIRPORTS BY CREATING

- A Validated Workforce through
  - Biometric authentication
  - Workforce management
- A Validated Facility through
  - Video monitoring
  - Driver/vehicle authentication
  - GPS devices to monitor vehicle traffic
  - Access control within aircraft, including the cockpit
- A Validated Communications Infrastructure through
  - Integrated communications infrastructure
  - Migration to a secure networked, digital technology
- Greater Protection and Respect for Passengers through Implementation of the Above Recommendations

AIR TRAVELERS MUST HAVE RELIEF 1
WITH RESPECT TO

- The Travel Environment Threatened by Inadequate Security
- Cumbersome Airport Internal Communications Systems
- Discomfort Resulting From Poorly Trained Personnel
- Lengthy Check-in Processes Due to Poorly Functioning Detection Technologies

MINETA SAN JOSE INTERNATIONAL AIRPORT IS IDEALLY SUITED FOR TESTING NEW TECHNOLOGIES

- Mid-sized
- Heavy Business Use
- Major Cargo Center
- Positioned for Redesign

NEW TECHNOLOGY APPLICATIONS MUST NOT INTERFERE WITH CIVIL LIBERTIES

- The Right to Privacy
- Freedom from Racial/Ethnic Profiling

EXECUTIVE SUMMARY

Much like the rest of American society, the commercial air travel industry is in the midst of profound change. The unprecedented airplane-launched attacks of September 11, 2001 by adversaries of democracy culminated a series of international incidents that tragically brought the horrors of terror and devastation to American soil. Clearly, our traditionally open ways of life will never be the same.

Fundamental questions emerged in the wake of the September 11 tragedy. Among them: What can we learn from the catastrophic events to make America and air travel safe again? What existing technologies utilized by Silicon Valley companies can be applied to the national air transportation system in general and to Mineta San José International Airport (SJC) in particular? How can we make our airports safer, yet preserve the individual liberties that distinguish the United States from many other countries? The Silicon Valley Blue Ribbon Task Force on Aviation Security and Technology Report responds to those questions with pro-active prescriptions based on novel, yet available applications of technology to the air transportation and airport environments.

The Blue Ribbon Task Force Report is the result of a 100-day intensive effort organized by United States Congressman Mike Honda and San José Mayor Ron Gonzales, led by Chairman John W. Thompson, and carried out by committed leaders in the Silicon Valley. It began with Congressman Honda and Mayor Gonzales calling upon technologists, entrepreneurs, civil rights experts and community leaders to examine various technologies which, if applied in the airport setting, would improve airport efficiency and passenger safety, while protecting constitutionally guaranteed individual rights. The report ends with a series of carefully tailored and targeted technology applications that offer workable improvements to the complete airport environment.

The report's recommendations focus neither upon specific "brand" names nor the use of any single technology. Rather, the Task Force has concentrated upon utilization of several technologies that, either alone or in combination, can make various aspects of the air travel experience safe once again, while preserving precious constitutional guarantees. In fact, every technology recommended by the Task Force has been done via the objectives of addressing the needs of passenger comfort and convenience, for without satisfied passengers, airport and air transportation simply can not thrive. Given the concentration of technology knowledge in Silicon Valley, the report recommends SJC as a testing ground for these problem-solving methods and applications. Further, the proximity of local expertise will allow adjustments and refinements to these applications as necessary, thereby facilitating replication elsewhere.

BACKGROUND

In many respects, America lost its innocence about any invulnerability to terrorism on September 11, 2001. Even more than the human death toll and physical property devastation, the terrorist attacks graphically illustrated a national security crisis of immense proportions. Never before had commercial airplanes been hijacked and used as weapons of mass destruction. Indeed, in the wake of September 11, the nation emerged with unprecedented concern for its collective safety as well as a pronounced anxiety about its future. The President urged the public to go about its "business as usual," yet the tragedies of that late summer day seemed to preclude anyone from doing anything close to normal.

Government agencies responded quickly to the September 11 terrorist events. They conducted exhaustive safety assessments of oil pipelines, nuclear power plants, water facilities and other sensitive installations. But few flash points distressed the nation's leaders and general public more than air travel, a transportation means through which nearly 700 million passenger enplanements occurred in calendar year 2000. Suddenly, travelers felt unsafe in airplanes; pilots worried about their ability to control their planes; airlines suffered deep financial losses; and airports seemed ill-equipped to protect their facilities, workforce and the traveling public from assaults on their lives and personal freedoms.

A PERVASIVE CLIMATE OF FEAR

3

The terrorist attacks rocked the long-standing assumption that somehow America was removed from the problems of the rest of the world. Collectively, the four airplane assaults claimed more than 3,000 lives and inflicted billions of dollars worth of property damage. The violent and well-orchestrated offensives ripped the psychological fabric of American society within hours of reaching their targets. When asked in a national survey days after the events about the extent to which the attacks had shaken their personal sense of safety and security, 63 percent replied a "great deal" or "a good amount," compared with 26 percent who answered "not too much."[1] At the same time, most of the respondents expected that their lives would not return to normal in the near term. According to the same national survey, 57 percent believed that the attacks would precipitate fundamental changes in the ways that Americans live their daily lives, compared with 39 percent who contended that the country would soon return to "business as usual."[2]

National leaders continued to caution the nation that the terrorist attacks upon Americans are likely to occur again. On February 6, 2002, Central Intelligence Agency Director George J. Tenet testified before Congress that Al Qaeda, the international terrorist organization suspected of the attacks on the U.S., had every capability to strike the U.S. again.[3] Director Tenet's remarks were repeated in an even more direct manner on May 20, 2002 by Federal Bureau of Investigation Director Robert Mueller. In a candid moment before Congress, Mueller stated, "There will be another terrorist attack [in the United States]. We will not be able to stop it."[4] Thus, more than eight months after the attacks, Americans and their leaders remain worried about further damage and loss of life.

[1] "U.S. Keen to Avenge Attacks," *Los Angeles Times*, September 16, 2001, pp. A1, A13.
[2] Ibid.
[3] "Qaeda Still Able to Strike the U.S., Head of C.I.A. Says," *The New York Times*, February 7, 2002, pp. A1, A10.
[4] "FBI Says U.S. Suicide Bomb Attack Likely, *Los Angeles Times*, May 21, 2002, p. A1.

A RAVAGED ECONOMY

The terrorist attacks of September 11, 2001 cut deeply into a national economy already faltering from recession. According to one financial report, the declines in economic activity reeling from the terrorist attacks subtracted an additional one percent from the annualized third-quarter gross domestic product growth beyond any reductions associated with the already slowing U.S. economy.[5] Another assessment projected a 16 percent decline in the value of the broad-based Standard & Poor's 500 stocks.[6] New York City alone suffered $83 billion in economic losses, including more than 130,000 jobs.[7]

The airlines and their support industries were hit especially hard. Air travel was shut down for four days after the attacks, costing the airlines $340 million per day in lost revenues.[8] When air service resumed, shattered passenger confidence left airplanes so empty that the airlines were forced to lay off nearly 150,000 employees.[9] Industry analysts predicted that the airlines would lose $6.5 billion in the twelve-month period following the terrorist attacks,[10] three times the losses previously expected in the recession-driven year, and a sharp reversal from the $2.6 billion profit earned during the previous twelve-month period.[11] Meanwhile, airline stocks tumbled by between 10 and 75 percent, with the average stock value reduced by 41 percent,[12] cutting a wide swath through the economic underpinnings of the American economy.

The ripple effects throughout the travel sector were even more serious. Boeing Corporation furloughed more than 30,000 workers in response to the decisions by airlines to reduce
4 their flying schedules by 20 percent or more, and to delay deliveries of new aircraft.[13] The Travel Industry Association projected nationwide losses of 453,000 travel/visitor-related jobs amounting to $43 billion.[14]

Airports, seen as benign way stations prior to September 11, were suddenly viewed as targets for weapons of mass destruction. Nowhere else in the country were the previous ways of doing business altered more dramatically and rapidly for more people than in air travel and airports. And now, with new minivan-sized explosive detection machines and massive conveyor belt systems required at airports, experts predict new unbudgeted airport retrofitting costs of more than $40 billion.[15]

5 Evan F. Koenig, "Down But Not Out: The U.S. Economy After September 11," a presentation to the Board of Directors [illegible] Federal Reserve Bank of Dallas, November 2001, p. 1.
6 David M. Blitzer, "September 2001 Trends and Projections—After September 11," Standard & Poor's, September 2001, [illegible]
7 "They'll take Manhattan," *U.S. News & World Report*, p. 34.
8 "Pork Barrel or Kick-Start?," *Newsweek*, October 15, 201, p. 66.
9 "Crisis Grows for Airlines Worldwide," *Los Angeles Times*, September 23, 2001, pp. A1, A16.
10 "Continental's Blunt Leader Faces Crisis, Again," *The New York Times*, September 23, 2001, p. C1.
11 "Aid for industry tanking anyway," *San Francisco Chronicle*, September 20, 2001, pp. B1, B3.
12 "Faith in flying key to airlines' survival," *San Jose Mercury News*, October 16, 2001, pp. 1C, 4C.
13 "What Kind of Rescue," *Business Week*, October 1, 2001, p. 36.
14 "Air Transport and the Bay Area Economy," Bay Area Economic Forum, January 2002, p. 4.
15 "Security upgrades strain airports' space, budgets," *USA Today*, May 29, 2002, p. 1.

Airline finances were already suffering prior to the September 11 events,[16] but the attacks ripped away any hope of normal cyclical economic recovery. Said U.S. Department of Transportation Secretary Norman Mineta, "We are [now] talking about the safety and security and the stability of an entire industry. Not of one or two or three or four companies, but an entire industry."[17] As a result of this seismic economic collapse, and fearing bankruptcy from the nation's airlines, Congress approved an emergency airline economic package with $5 billion in direct aid and another $10 billion loans within nine days of the attacks.

Within the Bay Area, airplane passenger and cargo traffic fell by more than 20 percent as a consequence of the terrorist activities and subsequent reactions. Hotel occupancy rates, having plunged to well below 50 percent immediately after September 11, remained at 60 to 65 percent of capacity as recently as January 2002.[18]

At SJC, major and normally lucrative long haul routes to Toronto, Taipei and Paris were dropped, costing not only the affected airlines dearly, but the local economy between $100 million and $200 million each on an annual basis. Additionally, airport concession revenues fell sharply from $9.3 million in October 2000 to $7.6 million in October 2001.[19] Parking revenue, which historically accounted for 46 percent of the airport's budget, plummeted from $38.7 million to $28.7 million between May 2001 and May 2002; at the same time, new security costs for the airport parking areas increased by $500,000.[20] Thus, SJC found itself with the unenviable combination of lower revenues and higher operating costs.

Whatever the direction of the U.S. economy prior to September 11, economic activity after that date declined at a much more precipitous rate than beforehand. And in tourist destination regions such as the Bay Area, the impact was more severe than elsewhere.

MAJOR ELEMENTS OF CONCERN RELATED TO AVIATION

The terrorist attacks rocked virtually all sectors of American security—governmental, corporate and individual. Within days of the horror-filled events, experts discussed U.S. vulnerability to biological warfare, chemical warfare and nuclear weapons. Each area of potential devastation presented its own unique set of troubling circumstances pointing to a serious absence of adequate public protection.[21] Nevertheless, the September 11 assaults were orchestrated via the seizure of commercial airplanes for use as lethal weapons. Given this method of aggression, several airport-related security issues emanated from the terrorist attacks and have remained worrisome to the general public. Chief concerns have focused upon aircraft/airport vulnerability, protecting civil liberties, implementing on-site airport security, and reductions in airport traffic.

[16] "Suddenly, Carriers Can't Get off the Ground," *Business Week*, September 3, 2001, pp. 36-37.
[17] "What Kind of Rescue," *Business Week*, October 1, 2001, p. 38.
[18] "Air Transport and the Bay Area Economy," *op. cit.*, p. 1
[19] *Ibid.*, p. 10.
[20] "S.J. Airport to Raise Hourly Parking Rate," *San Jose Mercury News*, May 30, 2002, pp. 1B, 6B.
[21] See "Terror Weapons: The Next Threat,?" Time, October 1, 2001, pp. 70-71, "Treating Terrorism ," *San Jose Mercury News*, October 21, 2001, pp. 1C, 4C, and "U.S. Nuclear Plan Sees New Targets and New Weapons," *The New York Times*, March 10, 2002, pp. 1A, 6A.

Vulnerability

The terrorist attacks of September 11, 2001 presented travelers and non-travelers alike with a new sense of vulnerability. No longer did travelers automatically assume a safe travel environment, whether in the airport or in the air. In addition, for the first time, collective apprehension spilled over into other transportation modes. Suddenly highways, rail and bus networks, inland waterways and ocean harbors became sites of great concern. Additional anxiety focused on sources of weapons materiel as well as targets; nuclear power plants and dams became part of a long list of facilities viewed by public authorities as focal points of possible attack, with the Environmental Protection Administration estimating as many as 15,000 chemical, water and waste treatment plants vulnerable to terrorist activities.[22] Governments invested precious scarce resources to protect the public and sensitive facilities.[23]

Yet, public anxiety remains. Six months after September 11th, 54 percent of the respondents in a nationwide poll agreed that the airlines needed to take additional steps to assure better security, compared with 38 percent who felt that the airlines had done enough. Sentiment toward the federal government's management of the anti-terrorism effort was a bit more favorable: 48 percent viewed the federal government as having done enough to promote airport security, with 45 percent stating otherwise.[24] Nevertheless, a national poll conducted by the *Washington Post* on May 18-19, 2002, found that only 46 percent of those interviewed believed that the government can stop future attacks, down from 55 percent who answered favorably in March 2002.[25] Clearly, these events have brought about widespread and long-lasting concern.

Civil Liberties

In the post-September 11 environment, the Department of Transportation has embarked upon a comprehensive security program of stricter scrutiny of passengers and their belongings. The additional security requirements at passenger checkpoints and more vigilant airline passenger assessments have produced long airport lines and bottlenecks for pedestrian traffic moving through the facilities. Along with the worsened airport physical conditions, security screeners have been criticized for poor training and inconsistent applications of mandated procedures.[26] Some passengers have been singled out and/or searched because of their physical appearances alone; others have been questioned by unnecessarily aggressive and poorly trained screening personnel.

22 "Bush wants terror risk evaluation for 15,000 plants," *San Francisco Chronicle*, June 8, 2002, p. A6.
23 "Additional Budget Cuts As States and Cities Address Safety Issues," *The New York Times*, November 15, 2001, p. B9.
24 "Skies Not Safe," CBS News Poll, March 14, 2002.
25 "President Retains Strong Support," *Washington Post*, May 21, 2002, p. A4.
26 Gerald L. Dillingham, "Aviation Security: Vulnerabilities in, and Alternatives for, Preboard Screening Security Ope[illegible] United States General Accounting Office, Testimony before the Committee on Governmental Affairs and Its Sub[illegible] Oversight of Governmental Management, Restructuring and the District of Columbia, U.S. Senate, September 25[illegible]

Questionable treatment of individuals going through the ticketing and/or screening process has raised its own distinct layer of serious constitutional issues. Civil liberties experts have focused upon the questions of privacy as protected by the Fourth Amendment, freedoms of speech and association as protected by the First Amendment, and due process as protected through the Fifth and Fourteenth Amendments. Indeed, for some travelers, the post-September 11 "war on terrorism" environment has elements eerily reminiscent of the hardships suffered by Japanese-American citizens at the onset of World War II.[27]

On-site Facilities Management

Commercial airports have been proven to be porous environments ripe for security intrusions. In one Department of Transportation study between December 1998 and April 1999, investigators breached airport security 117 out of 173 efforts, or a 68 percent success rate.[28] Between 1996 and 2000, federal undercover agents were able to sneak bomb-like devices beyond airline security personnel at SJC on nine separate occasions.[29] No doubt, such breakdowns contributed to the decision of Congress to direct replacement of virtually all airport passenger screeners with federal employees as part of the Aviation and Transportation Security Act.

These concerns became more pointed after September 11, when hijackers avoided detection at three airports prior to takeoff. According to the Federal Aviation Agency, between October 30, 2001 and April 6, 2002 alone, security lapses and breaches produced no fewer than 180 evacuations of air terminals, resulting in 540 flight cancellations and 1,923 flight delays.[30]

Because of new cumbersome airport security requirements, air travel has now become a time-consuming ordeal. Depending upon the airline and the airport, domestic passengers are now directed to arrive anywhere between 60 minutes and two and one-half hours before their flights, many of which do not last as long as the time required for airport processing. International passengers are required to arrive at least two to three hours before their flights. In addition, with new security measures allowing only ticketed passengers into the gate areas, people meeting arriving travelers are often inconvenienced by inadequate numbers of concession stands, restaurants and restrooms in non-secure sectors of the airports.

Reduced Air Traffic

Air traffic has been slow to recover from the September 11 attacks. As of February 2002, traffic at the nation's 31 largest airports was down more than 12 percent from the corresponding period in 2001.[31]

[27] "War on Terrorism Stirs Memory of Internment," *The New York Times*, September 24, 2001, p. A18.
[28] "How Safe Can We Get,?" *Time*, September 24, 2001, p. 87.
[29] "Hidden weapons slip by California security," Scripps-McClatchy Western Service, September 27, 2001. During the same period, agents penetrated security at San Francisco International Airport eight times and Los Angeles International Airport on six occasions.
[30] "Summer Fliers Likely to Face Endurance Test at the Airport," *The New York Times*, May 11, 2002, pp. A1, B2.
[31] "Air Traffic Is Off Almost Everywhere, but the Dip Is Uneven," *The New York Times*, March 12, 2002, p. A18.

Reduced passenger traffic has produced financial losses exceeding $7 billion in 2002 for the major airlines, about $4 billion more than they expected from the effects of the national recession.[32] Within a month of the terrorist attacks, airlines had furloughed more than 100,000 employees. Ripple effects were felt down stream from airport concession stands to airplane manufacturers, and beyond. At San José, airlines reduced the number of scheduled flights by 20 percent. Such losses have had deleterious impacts on tens of thousands of people in the local economy as well as the national economy.

Combined, these developments show little sign of abating in the near term. To that end, the Federal Aviation Administration predicts that air traffic by September 30, 2002 will remain 12 percent below the traffic of September 2001.[33] Such projections augur for continued airline losses and long-term layoffs for air industry and industry-related workers throughout the nation.

PASSAGE OF THE AVIATION AND TRANSPORTATION SECURITY ACT

On November 19, 2001, Congress passed the Aviation and Transportation Security Act (ATSA).[34] Enacted only ten weeks after the terrorist attacks, this legislation represented both a response of the United States government to international terrorism and a blueprint for national action to protect itself.

Comprehensive in scope, the ATSA contained several landmark provisions. The major
8 components of the 51-page legislative act emphasized aviation security and passenger safety, with the underlying objective of restoring passenger confidence in the U.S. air travel industry. Key elements of the legislation included:

- Creation of a new Transportation Security Administration within the U.S. Department of Transportation;
- Establishment of an improved and redesigned security system throughout the nation's commercial airports;
- Authority for the new Transportation Security Undersecretary to issue and oversee implementation of security-related rules and directives;
- One hundred percent baggage screening airport capabilities, with advanced explosive detection systems in place by the end of 2002;
- New airplane safety features such as fortified cockpit doors and other measures;
- A dramatically increased presence of sky marshals on many commercial aircraft;
- Development of new screening technologies;
- Replacement of 28,000 private sector-employed screening personnel with trained federal employees by the fall of 2002;

[32] *State of the U.S. Airline Industry: A Report on Recent Trends for U.S. Carriers, 2002*, op. cit., p. 3.
[33] "FAA Expects Fares to Decline This Year But Forecasts a Sharp Increase for 2003," *Wall Street Journal*, March 12
[34] Public Law 107-71.

- Full authority for the Transportation Security Undersecretary to hire and fire screeners as well as determining the conditions for their employment;
- Organization of a Transportation Security Board to review the actions of the Transportation Undersecretary

An important piece of the legislation provided opportunities to test "best practices" at no fewer than twenty pilot, or beta site airport facilities, with successful programs to be replicated and adopted elsewhere in the air transportation system. This amendment, originally co-authored by Congressman Honda and Congressman James Matheson in HR 3101, was folded into the ATSA.

Multifaceted in approach, the Aviation and Transportation Security Act covers a vast amount of public policy making territory. At its root, however, the new law seeks to improve aviation security and passenger safety, while restoring the confidence of the traveling public.

CRITICAL QUESTIONS

For all of its intentions and comprehensiveness, the ATSA left many critical questions without answers. Among the most important unanswered questions are:

1. What are the most vulnerable areas of airports with respect to the ability to effectively screen, identify and/or authenticate entry?
2. How can airports safely promote movement of materials and people within the physical facilities? 9
3. What new technologies might be utilized to make airport environments safe?
4. How can the Department of Transportation determine the highest return on technology investments, and what criteria would be necessary for such considerations?
5. How can the fundamental civil liberties of travelers be protected as the government moves to employ new methods to interdict weapons and agents of terror?

CREATION OF THE BLUE RIBBON TASK FORCE

With inadequate public resources to answer so many questions about air transportation safety and passenger security, Silicon Valley public officials searched for entrepreneurial and innovative solutions. One novel approach emerged with the creation of the Blue Ribbon Task Force on Aviation Security and Technology. Constructed out of a partnership between United States Congressman Mike Honda and San José Mayor Ron Gonzales, the Task Force was organized as a private sector/public sector collaborative effort to address the thorny issues of airport safety, security and convenience raised by the Aviation and Transportation Security Act.

Appreciating the reservoir of innovative technologies in the Silicon Valley, and sensitive to the trade-off between security and convenience, Honda and Gonzales sought out the knowledge and advice of technology experts. Thus, they designed a Task Force comprised of leading authorities with experience in technology, security, business, aviation and public policy.

At the first meeting of the Blue Ribbon Task Force, Congressman Honda and Mayor Gonzales challenged the members to search out and recommend the best technologies for responding to the issues of security, safety and comfort, while maintaining sensitivity to critical matters of passenger privacy. They appointed John W. Thompson, Chairman
10 and CEO of Symantec to serve as Chair. Recognizing the many ongoing concerns of an anxious public, Honda and Gonzales asked the Task Force to complete its work within 100 days.

TASK FORCE GOALS

The Blue Ribbon Task Force on Aviation Security and Technology dedicated its efforts to meeting four major goals:

1. To identify existing technologies that can be utilized to improve the security of the national air transportation system, with immediate deployment at SJC;
2. To identify and recommend how existing technologies can be combined into interoperable systems to improve airline efficiencies and customer convenience at airports;
3. To identify and recommend emerging security technologies and systems for development by both the federal government and private sectors at airports throughout the nation;
4. To design technology-driven recommendations in such a manner that they vigorou protect personal liberties of the passenger and workforce while promoting public saf

Guided by these objectives, the Task Force set out to examine air transportation-relat problems with the goal of finding appropriate technology-designed responses created companies in the Silicon Valley.

Governed by the 100-day mandate, the group limited its work to the following four general areas: improving security (including passenger and the cockpit), protecting individual civil liberties, strengthening workplace integrity, and improving customer service. Cargo was omitted from discussion because of pending TSA considerations. With this framework, the Blue Ribbon Task Force defined specific transportation-related problems and potential high technology solutions, bearing mind that technology is only as valuable as the judgment of those who use it. Ralph G. Tonseth, Director of Aviation for SJC, was asked to provide staff and any support services requested by the group.

APPLYING HIGH TECHNOLOGY TO SJC

Congressman Honda and Mayor Gonzales recommended SJC as an excellent location for technology applications. Located in the heart of Silicon Valley, the airport is close to many technology companies and easily accessible as a beta test or pilot site.

Approximately 13.1 million passengers used SJC in 2001, virtually unchanged from the prior year. The terrorist attacks severely curtailed travel in the final quarter of the year nationwide, but particularly at SJC. Through August 2001, passenger levels were 13 percent above the prior year. That overall levels for 2001 remained constant with the prior year shows the extent that SJC dipped in the fourth quarter.

SJC's passenger volume makes it the fourth busiest in the state of California behind Los Angeles (61.6 million as of 2001), San Francisco (34.6 million as of 2001), and San Diego (15.2 million). Other recent facts about SJC include:

- Commercial flights—average of 398 per day (2001)
- Cargo—300.0 million pounds (2001)
- Revenue—$106.0 million (est. FY 2001-02)
- Direct airport jobs—5,900 (2002)
- Number of local jobs generated—75,000 (2001)
- Business revenue generated from airport activity—$4.2 billion (1998)
- State and local taxes generated from airport activity—$471 million (1998)

SJC is a medium-size hub with a full range of services for both domestic and international travel. As such, the airport is an ideal beta site for experimentation. The location, size and complexity of the airport led Congressman Honda and Mayor Gonzales to view it as well-suited for the application and potential replication of new technologies.

THE PROCESS: SEARCHING OUT INDUSTRY AND PUBLIC PARTICIPATION

Silicon Valley has long operated with several unique axioms, one of which is that new commercial ideas often spring from unusual combinations of people, values and concepts. With that in mind, the Blue Ribbon Task Force solicited broad-based input by constructing an open participation process. Accordingly, the Task Force set up several opportunities for individuals and companies to communicate their recommendations so as to assure inclusion of the widest variety of ideas and technology-based applications. This process provided opportunities for continuous input from the earliest moments of the Task Force's organization to the point at which members considered and recommended various technology solutions.

Use of Committees

Given the very brief time frame in which to carry out its mandate, the Blue Ribbon Task Force created two committees to expedite the information gathering, member deliberations and recommendations processes in an efficient, yet fully vetted manner. The task force received progress reports from the committees on a regular basis and refined its thinking as necessary.

Technology Demonstration Committee—The Blue Ribbon Task Force organized a Technology Demonstration Committee to study, consider and recommend "best practices" technology applications. Established under the leadership of Beatriz Infante, CEO of Aspect Communications, the committee was divided into three subcommittees that focused upon technologies applicable to passenger safety and comfort, authenticating the airport workforce, securing the immediate airport environment and improving the flow of communications.

Proposal Development Committee—The Blue Ribbon Task Force also elected to form a second committee for vetting potential recommendations. Accordingly, a Proposal Development Committee was set up to consider and aggregate all materials and recommendations into a final package. Chaired by Michael Fox, Sr., Chairman of M.E. Fox & Company, the committee also assumed responsibility for weighing the various technology recommendations against fundamental societal values ranging from civil liberties to implementation practicality.

Each committee met on several occasions to carry out its assigned responsibilities. This execution of tasks via committees permitted the Blue Ribbon Task Force to pursue research, reviews and recommendations of new technologies quickly, while maintaining a "checks and balances" approach to protecting fundamental individual rights and civil liberties.

Public Hearings

In an effort to capture new ideas and applications from the widest varieties of sources, the Blue Ribbon Task Force announced a public hearing via a press release issued on April 5, 2001. Held on May 10, 2002 at the Silicon Valley Conference Center, the public hearing was designed to solicit input from individuals, organizations and/or companies that wished to comment on issues, problems and solutions relative to the charter of the task force.

Interested parties were invited to present their ideas in advance via the Blue Ribbon Task Force web site at www.sjcblueribbontaskforce.org, a listed telephone number or via mail to the San José Department of Aviation. Any participant who failed to sign up in advance was provided an opportunity to speak the day of the public hearing, which was open to the press (see Appendix H).

The Blue Ribbon Task Force public hearing drew twenty-nine speakers. Their comments ranged from discussions of biometrics applications to inadequate wages for airport passenger and baggage security screeners. A complete listing of public hearing speaker comments and materials is located in Appendix D.

Request for Information

Recognizing that various technologies might be beyond the collective wisdom and knowledge base of the Blue Ribbon Task Force, the Technology Demonstration Committee announced a Request for Information (RFI) for applications potentially suitable for demonstration in areas under examination. According to the procedures made public by the committee on April 5, 2002, applications would be accepted no later than May 10, 2002, at which time committee members would consider proposals for presentation and/or display.

A total of forty-one proposals were received by the May 10, 2002 cut-off date (see Appendix E). After reviewing the various proposals, the Technology Demonstration Committee recommended the demonstrations by six applicants with technologies addressing the needs as outlined by the committee at the scheduled Public Demonstration Meeting on May 31, 2002.

In addition to the RFI applications, the Technology Demonstration Committee considered selected high technology areas that did not receive sufficient public or corporate response to the RFI request. Based upon need in these essential areas, the committee members invited some "best practices" companies to present their technologies at the May 31, 2002 meeting, thereby filling a critical gap (see Appendix E).

Public Demonstration Meeting

On May 31, the Blue Ribbon Task Force sponsored the Public Demonstration Meeting at the Silicon Valley Conference Center. The demonstrations were divided into three categories:

Technology Demonstration Committee Selected Technologies

Technology Demonstration Committee Member Technologies

Technologies Solicited by the Technology Demonstration Committee

The full list of company names, spokespersons, addresses and technologies is included in Appendix F.

THE BLUE RIBBON TASK FORCE OPERATIONAL FRAMEWORK

Bearing in mind the organizational mandate to consider technology applications for problems related to airport workers, the airport workplace, the airport infrastructure and air passenger safety, the Blue Ribbon Task Force focused its research and recommendations upon those particular issue areas. As such, existing technology applications already under consideration or in various states of utilization remained outside the purview of the Task Force. In addition, the Task Force stayed away from operational solutions already managed by existing technologies. Accordingly, the Task Force developed a set of guiding principles, criteria for technology utilization, and a ranking formula for determining the desirability and applicability of high technology applications to the stated areas of need. These are discussed later in this report.

Guiding Principles

Prior to consideration of any specific technology or application, the Blue Ribbon Task Force developed a set of guiding principles for the Technology Demonstration Committee. Collectively, these principles became the bases upon which the committee extended recommendations for application to the Task Force.

1. The Blue Ribbon Task Force will examine technology applications dedicated to increasing security for the nation's commercial aviation system.
2. The Blue Ribbon Task Force will employ a proactive approach in considering solutions to the problems of commercial aviation security.
3. The Blue Ribbon Task Force will explore any technologies that will promote a predictable and satisfactory commercial aviation experience for all users.
4. The recommendation of the Blue Ribbon Task Force will span a security continuum that encourages the applications of various technologies in manners that significantly improve commercial aviation security, workplace safety and passenger comfort.
5. The Blue Ribbon Task Force recommendations will include appropriate sensitivity to compelling civil liberties and constitutional values, bearing in mind the importance of balancing the imperatives of individual rights and guarantees with the needs of collective security.

These five principles became critical guidelines for the activities of Blue Ribbon Task Force. Any recommendations put forth by the Task Force were filtered through these critical points before receiving task force approval.

Criteria for Technology Utilization

Determining bases of technology application was an essential objective of the Blue Ribbon Task Force. This responsibility was assigned to the Technology Demonstration Committee. Without such criteria, almost any idea would fit into the solutions matrix. After considerable review and discussion, the committee crafted the following criteria for technology utilization:

1. The recommended technologies will be accessible on a consistent (24/7) basis to those who use them.

2. The recommended passenger screening technologies will be dedicated to a passenger screening process that lasts no more than ten minutes.
3. The recommended technologies will be easily maintainable and user-friendly.
4. The recommended technologies will be networked and digital, rather than stand-alone or analog, thereby enabling "real time" access and sharing of information.
5. The recommended technologies will be standards-based and interoperable, thereby enabling future extensions and operation over wireless and wired networks.
6. The recommended technologies will not jeopardize individual rights and liberties.

Ranking Matrix

In addition to developing criteria for the utilization of various technologies, the Technology Demonstration Committee established a list of factors used to determine the suitability of those technologies. The five factors are as follows:

Security—the more the technology enhances security in the workplace among employees or among passengers, the more desirable it is; the extent to which the new technology does not improve either workplace or passenger security makes it less desirable.

Cost of the technology application—The lower the cost, the more desirable the proposed technology application; the higher the cost, the less desirable the proposed technology application. 15

Maturity of the technology—The more mature the technology, the less the risk of failure; the less mature the technology, the greater the risk of failure.

Time to deployment—the more quickly the technology can be deployed, the more desirable; the less quickly the technology can be deployed, the less desirable. The extent to which the technology can stand alone without connecting to a national network further enhances its value.

Intrusiveness—to the extent that the technology does not intrude upon civil rights and/or civil liberties, it is more valuable; to the extent that the new technology intrudes upon civil rights and/or civil liberties, it is less valuable. The committee strongly believes that applications of all technologies must be done in ways that are scalable and rules-based, assuring fairness to the greatest extent possible.

Few technologies considered by the committee are exclusively low cost, mature, immediately deployable, completely secure and totally non-intrusive. Nevertheless, the closer a technology comes to these ideal types, the more likely that the technology would receive a strong committee recommendation.

Collectively, the guiding principles, the criteria for technology utilization, and the ranking factors became the bases upon which the Technology Demonstration Committee made its recommendations to the Blue Ribbon Task Force at a meeting on June 4, 2002.

OBJECTIVES, PROBLEMS AND RECOMMENDED TECHNOLOGY SOLUTIONS

The Blue Ribbon Task Force placed passenger security, comfort, protection, and integrity at the forefront of its work. These boilerplate values were critical to the full complement of efforts carried out by the group. From the earliest research statements to the point of recommendations, the Task Force considered technologies that ultimately would contribute to a more efficient and safer air transportation environment without sacrificing any individual passenger liberties.

The structure and operations of commercial airports pose significant security and safety challenges. Airports are composed of multiple functional areas, including commercial aviation terminals, general aviation terminals, and cargo and freight operations. Each area has domains requiring different security levels. For example, in the commercial aviation area, there is an unsecure terminal area, a secure terminal area, and an "air-side" area which includes the jetway and ramp areas. There are not always "hard" boundaries or controlled access points between these domains and the functional areas where security functions such as authentication and inspection can be reliably enforced. In addition, personnel associated with several different commercial companies and government organizations may require legitimate access to one or more of these areas. These individuals are usually involved with the movement of a wide variety of materials including baggage, food, fuel, and other cargo and, as such, must be validated as they move into and around the airport facility to carry out their tasks.

Compounding these challenges is that no one organization has responsibility for all airport security functions. Federal agencies, including the FAA, TSA and FBI, local law enforcement, airport security operations and airline personnel all are involved with various aspects of airport and aviation security. Under these circumstances, there can be jurisdictional gaps, leading to security breakdowns and significant challenges in real-time coordination of activities during a crisis.

In examining the flow of people and materials through the entire airport venue, the Blue Ribbon Task Force determined that the entry point protection offers the best opportunity for preserving safety and security throughout the process. Thus, the Task Force concluded:

1. It is vital to secure the workplace environment as employees enter sensitive facilities;
2. It is imperative to authenticate employees before they go into the workplace;
3. It is critical to maintain state-of-the-industry information exchange procedures at all levels and conditions of communications;
4. It is essential to safeguard passenger rights from the earliest entry points on and throughout the traveling experience.

After reviewing several issues and approaches for solutions, the Blue Ribbon Task Force narrowed its activities to four areas: workforce security, facility integrity, the airport communications network, and preservation of passenger dignity. Each area of examination contains an objective, a problem and recommended high technology-based solutions. Low tech (e.g., more telephones or personnel), operational solutions (e.g., reducing the number of entry areas to control access or personnel training) and airport governance and general management were not considered because airport administrations and various other agencies are addressing these issues. In addition, the Task Force has refrained from considering solutions relative to the movement of baggage because of existing Transportation Security Administration activities and the technology applications already exist for validating baggage and are in various stages of implementation.[35]

The Task Force recognizes that the private sector and several government agencies have already moved in the post-September 11 era to expand research and intelligence gathering efforts. Such endeavors are valuable first steps toward a safer airport environment. To that end, the Task Force views its technology-driven proposals as the next step in what must be viewed as a long process of technology research, development and application.

The discussion below focuses upon recommended "best practices" for the four discussion areas. Recommendations are determined as a result of presenting an objective, identifying the problem, proposing solutions and describing passenger benefits. In some cases, the same technology is recommended for more than one area; in other instances, combinations of technologies are recommended. The discussion assesses the viability of each proposal
in terms of the ranking factors discussed above as "high," "medium," or "low."[36] The . 17
complete list and analyses of possible technologies presented at the May 31 meeting is found in Appendix F.

VALIDATED WORKFORCE SECURITY

The Blue Ribbon Task Force contends that a trustworthy workforce is the cornerstone for a safe air transportation environment. Congress has taken the first step of replacing private security screening personnel with government employees, thereby addressing one aspect of a secure air transportation network workforce. Nevertheless, the totality of the workforce in the airport setting extends well beyond screening employees to individuals who work behind the ticket counters, in concession venues, on the tarmac, in supply vehicles and anywhere else with direct or indirect access to any elements of the air transportation system.

Objective

To assure that all elements of the extended airport workforce, especially those connected with the Security Identification Display Area (SIDA), consistently satisfy the highest possible security standards.

[35] "U.S. Transportation Secretary Mineta Announces Successful Test of New Technology to Secure Cargo Movement in U.S. Ports," press release, June 4, 2002.
[36] With respect to the deployment category, the assessments are "Easy," "Medium," and "Hard."

Problem

Control of the airport workforce, especially in the Security Identification Display Area (SIDA), is weak due to inconsistent standards, uneven oversight, poor enforcement and multiple constituencies that range from local to national authorities. Accordingly, opportunities exist for unauthorized individuals to compromise the integrity of the workforce through the use of false identification, unauthorized presence in authorized areas and "piggybacking," an entry process that allows for the possibility of an unauthorized employee to quickly move behind an authorized employee through a doorway entry.

Technology Solution #1: Biometric Authentication

Biometric authentication mechanisms should be utilized for identifying all employees who require access to airport functional areas, especially the SIDA. Reliable authentication is based on combining at least two of the following identification factors: something an individual knows (password, PIN etc.), something one has (an ID card), and an individual's unique characteristic (fingerprint, iris scan, etc.). An identification card with biometric information combines the latter two factors and is an extraordinarily reliable authentication mechanism. A variety of biometric characteristics can be used to establish identity, including fingerprints, facial scans, iris scans and hand scans. In general, the costs of storing and scanning most biometric characteristics pale next to the general costs of performing the initial identification for card issuance and the costs of administering the infrastructure to manage and revoke cards once issued.

Any identification card should encode all information, including biometric data, so that it is electronically scanable. Several technologies are available to carry out this activity including magnetic stripe, RF, optical, and "smart card" which contains a small microcircuit or chip. Irrespective of encoding technology, all identification cards must include a mechanism to authenticate the card itself and its information, preferably through the use of digital certificate technology. This mechanism also enables one of the most important attributes of an effective authentication system—the means to revoke or cancel a validated card.

Biometric scanning devices should be networked at access control points where biometric data is collected from an individual and compared with biometric data on the card. This enables the scanning device to validate the identification card, ensure currency, and determine that it has not been flagged or revoked. It also enables individuals to be located on a "real-time" basis, which is vital in many security-related scenarios. Biometric-based access control mechanisms should be augmented with some form of monitoring either by co-located individuals (security personnel) or by remote personnel using video technology.

Ranking Matrix

Security:	High
Cost:	Medium
Maturity:	High
Deployment:	Easy
Intrusiveness:	Low

Technology Solution #2: Workforce Management

A critical aspect of workforce security lies with the ability to schedule, track and monitor employees. Although the federal government is assuming responsibility for security personnel, local airport employers will retain control of workforce activities and access. Such controls are necessary to ensure that the right person is at the right job at the right time. In order to promote a secure workforce, the Blue Ribbon Task Force recommends workforce management software to automate scheduling, skills management and access control. To the extent that an employee is detected in a wrong area based on schedule, skills and/or access control, the real-time alerts and notifications engine will send a message to the appropriate security personnel. Additionally, changes in schedules can be controlled in real-time to allow for the shifting airport environment.

Ranking Matrix

Security:	High
Cost:	Medium
Maturity:	High
Deployment:	Easy
Intrusiveness:	Low

Passenger Benefits

The most important source of reassurance to the passenger lies with the firm belief that authorized personnel are where they belong and that supervisory authorities have fast, accurate methods to detect any situation contrary to that expectation. The knowledge that employees are not compromised by impostors will go far toward restoring passenger confidence in the airport environment and flying experience.

VALIDATED FACILITY

A validated facility is the backbone of the airport's physical environment. Within a validated facility, movements of people and materials are monitored, therefore guaranteeing that the materials within the area belong there and that unauthorized goods or individuals will not enter the area. The activity within the validated facility is "sterile" to the extent that it is not compromised by materials or individuals entering the facility without approval from an appropriate authority.

Objective

To provide for a secure airport facility, especially the Security Identification Display Area (SIDA) in which the movement of employees, vehicles and baggage take place without compromise from outside, tainted or unauthorized sources.

Problem

Validating or securing the airport facility is a daunting challenge for several reasons: First, the facility includes multiple functions and areas. Second, various personnel belonging to several different commercial companies and government organizations require access to those areas. Third, a variety of materials including baggage, food, fuel, and other cargo must be moved into and around the airport facility. For the facility to be secure, individuals must be subject to authentication and access control as they move between areas; likewise, trusted authorities must be able to inspect and/or validate materials coming and leaving the facility without tampering.

Technology Solution #1: Video Monitoring

It is recommended that airports significantly expand and migrate their use of video monitoring technologies. Video monitoring both enables more effective security operations and reduces security costs. Generally, remote video monitoring from a central location is a more effective and less costly enterprise than on-location patrols by security personnel. Video technology provides a "force multiplication" factor by enabling security personnel to monitor many areas across a spread out site. Given the extensive perimeter, multiple access points, and multiple interior areas that require surveillance, video monitoring can be instrumental in securing airport facilities. Along with airport security management, extensive video monitoring also protects general health and safety within the airport facility by watching for fire or accident.

Although most airport video monitoring is based on analog (CCTV) technology, the Blue Ribbon Task Force recommends aggressive migration to digital video technology because of its numerous benefits. For example, in many security scenarios, high-resolution digital imaging can capture facial characteristics at a distance. Digital video also enables the use of a common and standard networking infrastructure rather than requiring a dedicated, stand-alone set of cabling, making it more cost-effective and more extensible than analog video. In addition, by coordinating with other forms of digital data, digital video enables "real-time" correlation from an access control point with information from a biometric identification card used for authentication. Digital video also enables more convenient and effective storage of images and enables rapid access of images based on a variety of selection criteria. Finally, digital video provides convenient "real-time" sharing of images to the various security personnel and organizations involved with airport and aviation security.

Ranking Matrix:

Security:	High
Cost:	High/Medium
Maturity:	Medium
Deployment:	Medium
Intrusiveness:	Low/Medium

Technology Solution #2: Driver/Vehicle Authentication

The Blue Ribbon Task Force recommends three steps to authenticate a driver and the [illegible] vehicle there prior to entering the airport facility. First, Driver Authentication—All [illegible] employees with access to vehicles seeking entry to the airport facility first must be [illegible] authenticated using the Biometrics process outline described in the Validated Work[illegible] Technology Solutions # 1 Biometric Authentication. Second, Vehicle Inspection—[illegible] vehicle should undergo a visual inspection and manifest/load comparison. Third, [illegible] Seal—Where appropriate, vehicle loads should be sealed to prevent tampering. [illegible] driver and vehicle have been validated, the task force recommends placement of [illegible] inspection "certificate" in the vehicle. This certificate should contain a GPS [illegible] so that vehicle movement can be tracked within the facility and in nearby areas [illegible] safe aviation operations such as freight forwarders, goods delivery companies [illegible] utilities and securities services. (See Technology Solution # 3, GPS Devices [illegible] Vehicle Traffic).

Ranking Matrix

Security:	High
Cost:	Medium
Maturity:	High
Time to Deployment:	Moderate
Intrusiveness:	Low

Technology Solution #3: GPS Devices to Monitor Vehicle Traffic

Given the movements of large numbers of vehicles at and near the SIDA, placement of global positioning devices in each vehicle authorized for airport access will allow authorities and managers to monitor and track traffic as it takes place. In order to maximize efficiency, the GPS could be associated with the authentication process at the entry and exit points, and other sensitive areas cited in Technology Solution #2. This "real time" management will permit immediate interdiction of unauthorized vehicles or authorized vehicles that travel to unauthorized locations.

Ranking Matrix

Security:	High
Cost:	Medium
Maturity:	High
Deployment:	Easy
Intrusiveness:	Low

Technology Solution #4: Access Control Within Aircraft

A biometric device system is recommended for pilots. This technology solution requires a nationwide identification system previously recommended in the Workforce Facility section. The Task Force recommends access control points within aircraft that will utilize the biometric authentication mechanism. As a result, access to sensitive areas will be limited to validated aircraft personnel including pilots, flight attendants, maintenance workers and other authorized personnel.

Ranking Matrix

Security:	High
Cost:	Medium
Maturity:	High
Deployment:	High
Intrusiveness:	Low

Passenger Benefits

With materials, employees, aircraft and vehicles closely monitored throughout the air transportation process at and near the airport facility, the SIDA will be secure and safe from unauthorized entry. Such an environment will assure passengers of a secure airport facility, authorized personnel and airplanes free of any tampering.

VALIDATED COMMUNICATIONS INFRASTRUCTURE

Personnel representing the multiple organizations responsible for airport and aviation security need to communicate in a seamless fashion at all times, but particularly during crisis. Whether within the airport, between the airport and the outside environment, or between the airport and non-airport authorities in other jurisdictions, exchanges of information and data must be rapid, accurate, and secure to assure efficient operation. The communications infrastructure must insure the integrity of information and must also prevent access or intercept by unauthorized personnel.

Use of a common network infrastructure across all organizations in the airport community and integrating multiple applications would greatly reduce operational costs, enable rapid extensions, increase resiliency, and promote information sharing from multiple sources in "real-time." In addition, all devices that perform vital security operations should be networked; they include, but are not limited to, video cameras, biometric scanning stations and baggage scanning systems. Networking enables information collected by these devices to be accessible on a "real-time" basis and shared with appropriate security organizations and personnel. In addition, networking instantly avails information about the status of devices. There have been many cases where significant disruptions to local airport operations and the nationwide aviation grid could have been avoided, had security personnel been instantly informed about a dysfunctional scanning system such as the loss of power.

22 ***Objective***

To provide a communications infrastructure within the airport and beyond that guarantees secure, real time transmission of data and other information, thereby assuring interoperability between technologies, immediate response capabilities and integration with legacy systems.

Problem

Existing airport communications infrastructures are generally composed of multiple, application-specific networks. In some airports, over fifty different, unconnected networks exist for various voice, video and data applications, with many duplicated for the multiple commercial and government organizations in the airport community. Each network has its own administrative costs as well as scalability and extensibility limitations. Moving all applications for every involved organization on to a common, standards-based and extensible network infrastructure would represent significant savings of operations costs. Such a change would also simplify and speed enhancements and extensions, increase resiliency, and most importantly, enable "real-time" information sharing.

Technology Solution #1: Integrated Communications Infrastructure

Typical airport communications infrastructure consists of multiple stand-alone, propri[illegible] application-specific networks. These networks are high in cost and difficult to admin[illegible] Typically, they do not allow for growth or change. An integrated digital communic[illegible] infrastructure would provide real time communications, data sharing and enhanced [illegible] security. In addition, an integrated communications infrastructure would allow mu[illegible] organizations (both on and off the airport site) to share critical information in real[illegible] The communications infrastructure would take advantage of open standards to en[illegible] future growth and flexibility. This massive infrastructure upgrade is the cornersto[illegible] security system.

Ranking Matrix

Security:	High
Cost:	Medium
Maturity:	High
Deployment:	Moderate
Intrusiveness:	Low

Technology #2: Migration to Networked, Digital Technology

All devices performing vital security operations should be networked, including video cameras, biometric scanning stations and baggage scanning systems. A standards-based digital communications technology is fundamental to the ability of devices to utilize a common network infrastructure, including both wired and wireless connectivity. To accommodate growth in high bandwidth applications, including digital video, the airport communications infrastructure will need to leverage the high bandwidth capabilities of wired network connectivity based on optical and 10 Gigabit Ethernet technologies. To accommodate remote, untethered, and mobile applications, the airport communications infrastructure will need to leverage wireless network connectivity based on cellular and 802.11 technologies. Use of digital technology is also fundamental to enabling the convenient storage, access, correlation, and sharing of the data collected by each device. Digital communications technologies compatible with standards-based Internet connectivity architectures are essential to enabling a scalable, extensible, and interoperable communications infrastructure.

The airport information technology (IT) network infrastructure should conform to standard security best-practices and architectural principles to ensure cyber security. These include compartmentalization of network domains to limit propagation of attacks and "virus infections." The airport information systems security officer (ISSO) should develop a defense in depth architecture to incorporate security technologies at all levels of the network, including the gateways. This includes security devices such as firewalls, network and host-based intrusion detection, internet and email content filtering, and anti-virus technologies all managed by a single central security management system. All networked devices should provide device authentication, and all communications between networked devices should be encrypted using virtual private network (VPN) technology. Periodic evaluations of IT operating system vulnerabilities should be conducted via automated assessment tools and by running vulnerability scans to assure that systems are properly patched and operating at peak security efficiency.

Ranking Matrix

Security:	High
Cost:	High/Medium
Maturity:	High
Time to Deployment:	High/Medium
Intrusiveness:	Low

Passenger Benefit

The ability of airport personnel to exchange information fully and in a secure manner will provide a strong protective layer of passenger safety. With fast, integrated communications throughout the airport facility, passengers will be safeguarded from unauthorized intrusions or unanticipated disruptive events, thereby ensuring the security and integrity of passenger and employee databases and significantly increasing a collective sense of safety and overall comfort. These databases should be monitored for evidence of tampering by conducting automated periodic security posture assessments of the database.

VALIDATED PASSENGER

Of all the elements associated with the airport environment, no one has suffered more than the passenger—an irony, given that passengers are the lifeblood of the flying experience. Subjected to long security waits in line, inconsistent racial and ethnic profiling and/or repeated body searches and other forms of individual intrusion, passengers have seen their dignity reduced to an afterthought. The passenger can be protected by a safe facility, a well-trained and screened workforce and an efficient communications network. Collectively, the technology-driven solutions will enable passengers to move people through the entry and boarding processes quickly and within an environment that honors constitutionally guaranteed freedoms.

BALANCING THE INTERESTS OF THE GOVERNMENT AND THE INTERESTS OF THE INDIVIDUAL

The recommendations above reflect the potential of technology for ensuring security. But in a free society, technology does not operate in a social vacuum; rather, it interacts with legal and ethical principles that underscore core values. Accordingly, the Task Force recognizes that recommended technology-based solutions must balance security needs with the values of privacy and other civil liberties.

Airports are quasi-public entities.[37] Although they include private organizations such as airlines and concessionaires, airports operate under the ultimate control and supervision of the government. Thus, both private sector operations and public sector guarantees co-exist and in some cases overlap within the general airport boundaries. Airline employees, vendors and shopkeepers are hired and fired in accordance with the rules of their companies; however, the traveling public is protected by Constitutional guarantees. Because of the public constituency and the use of public police powers, governance at airports is comparable to the governance of bridges, roads or any other enterprise which, as their underlying bases of existence, serve the public good.

Technology, the Private Sector and Constitutional Issues

Although everyone is protected by the U.S. Constitution, people who work for the [illegible] sector do so as part of a contractual relationship between the company and the [illegible] Some rules affecting that relationship are legislatively established—the minimum [illegible] health and safety conditions, and anti-discrimination provisions are examples. N[illegible] in exchange for working in a secure environment, employees assume certain resp[illegible] that may subject them to more security and validation. Thus, the application [illegible] or other monitoring means to employees and their workplace strikes an app[illegible]

[37] *Cf.* United States v. Davis, 482 F.2d 893 (9th Cir. 1973) (concluding that an airport ticket agent's search [illegible] was "part of the overall, nationwide anti-hijacking effort, and constituted 'state actions' for the purposes of [illegible]

balance, assuming that biometric and private information is safeguarded. Furthermore, real time video streaming, GPS placement or other means of identifying location, authenticity or propriety of activity are usually well beyond the reach of right to privacy issues among employees.

Technology, the Public Good and Constitutional Issues

Using technology applications to solve security problems as they relate to passengers creates a set of issues distinctly different from those relating to employees or the workplace setting. Inasmuch as it is reasonable to view passengers as members of the public and the airport as a public place, the threshold for violating individual rights takes on a higher standard altogether. Yet, terrorists can pose as passengers, thereby creating a civil liberty versus security conundrum. This tension has led the Transportation Security Administration to consider new methods of passenger identification and scrutiny that may redefine passenger safety and civil rights.[38]

Recent public opinion data underscore anxieties about the passenger security issue. In a national poll conducted by the *New York Times* and CBS News in December 2001, 64 percent of the respondents agreed that it is a "good idea" to "make changes in the [civil] rights guaranteed by the Constitution." Yet, in the same survey, 55 percent were "very concerned" or "somewhat" concerned about losing some of their civil rights.[39] This is the haystack through which the high technology needle must thread new security applications while not intruding unnecessarily upon the rights of individuals.

25

The constitutional rights and values most at risk from high technology applications to airport-related security are those associated with the right to privacy (Fourth Amendment), freedom of expressions and association (First Amendment) and due process of the law (Fifth and Fourteenth Amendments). Nevertheless, the Blue Ribbon Task Force believes that its recommendations will not adversely impact airport passengers.

The Fourth Amendment: the Right to Privacy

The Fourth Amendment regulates how and when authorities may engage in the search or seizure of a person or property. Under ideal circumstances, police ask a judge for a warrant prior to undertaking a specified search or seizure activity. Warrants generally are not required when a search or seizure is conducted in a regulated environment such as an airport.[40] Nonetheless, because the long-revered value of privacy enshrined in the Fourth Amendment protects individual dignity and autonomy, technologies should be used in a manner that does not impinge upon privacy or unreasonably singles out some individuals over others. Much of the right to privacy debate now centers upon the extent to which the United States is on a wartime footing[41]—a condition that could impact the way the courts and other public institutions view the privacy issue in public settings such as airports. The recommendations of the Task Force are directed primarily toward safeguarding the workplace, screening employees, and improving communications. Passengers are not the focus of the Task Force's proposals, although they are very much the beneficiaries. Moreover, none of the recommended technology applications have the potential (as face recognition

[38] "Plan Sharply Tightens Airport Screening," *The New York Times*, May 30, 2002.
[39] "Public is Wary But Supportive On Civil Rights Curbs," *The New York Times*, December 12, 2001, pp. A1, B9.
[40] See U.S. v. Bulacan, 156 F.3d 963, 967 (9th Cir. 1998) ("searches conducted as part of a general regulatory scheme, done in furtherance of administrative goals rather than to secure evidence of a crime, may be permissible under the Fourth Amendment without a particularized showing of probable cause").
[41] "Civil Liberty vs. Security: Finding a Wartime Balance," *The New York Times*, October 18, 2001, pp. A1, B6.

software would, for example) to radically change the amount of private information that airports, airlines, or the government gathers about the public. The Task Force is confident about safeguarding individual rights in a public setting.

The First Amendment: Freedom of Speech and Association

Freedom of speech is a basic right protected by the First Amendment. The right of association, while not explicitly mentioned in the Constitution, also has been viewed by the United States Supreme Court for nearly fifty years as a fundamental right related to advancing one's beliefs and ideas.[42] The rights of both free speech and association could be threatened with the universal applications of some technologies in the airport environment. For example, a sensor system that singles out individuals on the basis of what they say while traveling through the screening area could have a chilling effect upon free speech. Does this mean that such technologies should not be used? Not necessarily, if they are applied to the voluntary setting of the workplace. But those law enforcement personnel who elect to utilize technologies that potentially impact speech need to realize that they are jeopardizing protection of a precious right if they extend use beyond voluntary relationships. As such, they need to apply electronic monitoring only on an ad hoc basis and to the extent it corroborates other information.

The Fifth and Fourteenth Amendments: Due Process

Modern due process doctrine focuses upon the procedures used by government entities to determine whether a person should be subject to a particular legal restriction or requirement.[43] Cased in the Fifth and Fourteenth Amendment, the due process clause attempts to balance the interests of government with those of the individual, for example the individual's right to travel.[44] Due process remains a critical constitutional guarantee in the wake of September 11. More than ever, the government must carefully weigh the conditions under which it extracts information from individuals under suspicion especially when using seemingly removed, yet invasive means such as sensors or other data collection technologies. Balancing the need for security against privacy of the individual, therefore, requires technology deployment that is reasonably effective and accurate, while providing for due process protections.

Preserving the Balance

Maintaining the balance between security and constitutional values demands constant vigilance and attention to the details of how technology will be used and who will be affected. Aviation security and safety presents one setting which is deserving of such attention. The debate over how and the extent to which the federal government should increase passenger screening methods now sits at center stage in the this setting. It may well be that if the Transportation Security Administration adopts a computerized passengers profiling system (CAPPS), civil libertarians will challenge the new policy as a violation of privacy, particularly if such a system is mandatory. Equally unclear is whether such a plan will withstand judicial or legislative scrutiny. Whatever that outcome, no such proposal is forthcoming from the Blue Ribbon Task Force.

[42] NAACP v. Alabama, 357 U.S. 449 (1958).
[43] Brock v. Roadway Express, Inc. 481 U.S. 252 (1987).
[44] See United States v. Laub, 385 U.S. 475, 481 (1967).

TOWARD THE FUTURE: UNRESOLVED ISSUES

The technologies recommended by the Blue Ribbon Task Force represent only the first steps toward making airports trusted and secure 21st Century venues of commerce and travel. Nevertheless, as the Blue Ribbon Task Force conducted its research, members happened across other issues and solutions which, if applied with technology applications, would go far toward making the airport a safer, more secure, user-friendly environment. Accordingly, the Task Force recommends that authorities investigate and respond to the following:

Reengineering of Physical Space

Airport authorities may wish to consider different uses of physical space. With fewer people allowed past the passenger entry areas, space at and near the airline counter areas is at a premium. Individuals dropping off or picking up passengers have literally no place to go, contributing to highly congested outer lobbies. The accumulation of large numbers of people in addition to travelers moving through the process leaves the outer areas anything but user-friendly, revealing considerable frustration and possibly presenting additional security issues. In addition, limited space in the outer lobby area allows for very few concessions and other facilities.

Customer Service Training

Most airports tend to operate as an awkward, disjointed collaborative between public employees, the airport and other miscellaneous service personnel. In some cases, tasks among the various groups overlap; in other cases, there are gaping holes and conflicting rules of conduct. The Blue Ribbon Task Force recommends that the airport embark upon a training program for all employees regardless of employer or responsibility. The program would provide employees with basic information to pass on to inquiring travelers.

An Anti-Theft Program

Passenger screening often leaves the individual separated from his/her carry-on bags for several minutes, an anxiety-provoking situation that only worsens if an individual is singled out for further screening. Regardless of how fast people move through the screening process, the Task Force recommends that the Transportation Security Administration hire individuals responsible for nothing other than making sure that individuals are matched with their carry-on bags.

Registered Passenger Cards

Use of a registered passenger card might speed up the process of entering the airport secure area. Such a card could be purchased, the funds from which would be used to conduct research on the cardholder's background, arrest record and other valuable pieces of data. An individual with a registered passenger card would simply swipe it at an early checkpoint. This does not suggest, however, that the cardholder would be exempt from security screening; rather, the principal benefit would be early movement through the security line. If pursued on a voluntary basis, registered passenger cards might free up valuable, congested space, thereby allowing all passengers to move at a quicker pace, although it is not clear whether such a program would have significant security and efficiency benefits. The committee recognizes possible constitutional concerns with this idea.

General Aviation Concerns

Left without discussion by the Blue Ribbon Task Force is the issue of general aviation. Privately owned planes and private charters often operate out of small terminals with few, if any security-screening equipment. The numbers are compelling. With more than 200,000 small, privately owned airplanes operating in over 18,000 airports, there are endless opportunities for security breaches.[45] Almost all of these airports are outside of TSA control, although more security features exist in the instances where general aviation services share facilities with the commercial side, as in the case of SJC. With increased passenger frustration at traditional commercial airports, passengers may be more willing to book charter flights at relatively security-free general aviation airports, thus adding both to traffic and concern.

Technology Research Agenda

The Blue Ribbon Task Force worries that the federal government may not be structured for careful examination of transportation security. In this report, the task force has carefully culled the best current technologies available for application now. But what happens in a month, or in six months, or in a year when newer technologies come on line? What research should be initiated in order to fill technology gaps? The federal government is the level of authority best designed to answer this question. To that end, the task force hopes that the Department of Transportation or other appropriate agency will have available an adequately staffed office charged with the tasks of scrutinizing, analyzing and recommending new technologies for application to the airport setting on an ongoing basis.

Cargo

Although the U.S. Department of Transportation has taken steps to protect cargo containers, the Blue Ribbon Task Force expresses concern that the department has not attended sufficiently to air cargo, particularly to the extent that it relates to mail. Under current post-September 11 regulations, air carriers are forbidden from accepting mail parcels of more than sixteen ounces, thereby cutting deeply into the airlines' revenue stream. While the Task Force did not deal with this issue, the members encourage DOT to further explore ways in which technology can overcome this problem.[46]

Transport Workers Identification Card

One approach to employee authentication may be through use of a Transport Workers Identification Card. This card, currently under study by the DOT, would provide universal identification and verification for all transport workers within an airport and throughout all airports. In addition to matching personal data with the holder, the TWIC could also be embedded with biometric information such as finger prints, hand geometry, facial and dental structure or iris shape. The Task Force recognizes the presence of sensitive privacy issues here, but nonetheless believes that the potential benefits of this proposal suggest the need of further research.

[45] "Private Plane Charters: One Way Around Air Security," *The Washington Post*, June 2, 2002, pp. A1, A7.
[46] The Task Force appreciates that the TSA has devoted some attention to this important issue. See "Terror Risk Cit[illegible] Carried On Passenger Jets," *The Washington Post*, June 10, 2002.

APPENDIX A

Blue Ribbon Task Force Participants

A-1

Appendix A – Blue Ribbon Task Force Participants

Last Name	First Name	Position	Organization
Bold = Task Force Members			
Honda	**Mike**	Congressman	Member, U.S. House of Representatives (California 15)
Van der Heide	Jennifer	Chief of Staff	Office of Congressman Mike Honda (CA-15)
Maben	Meri	District Director	Office of Congressman Mike Honda (CA-15)
Mitchell	Chris	Legislative Assistant	Office of Congressman Mike Honda (CA-15)
Bostick	Matt	Field Representative	Office of Congressman Mike Honda (CA-15)
Gonzales	**Ron**	Mayor	City of San José
Webb	Jim	Senior Policy Advisor	Office of Mayor Ron Gonzales
Vossbrink	David	Communications Director	Office of Mayor Ron Gonzales
Chair:			
Thompson	**John W.**	CEO & Chairman	Symantec Corporation
Rak	Adam	Manager, Government Relations	Symantec Corporation
Paden	Cris	Public Relations Manager Corporate Communications	Symantec Corporation
Araki	**Sam M.**	Chairman	Security Technology Ventures
Ashby	**Dan B.**	Captain	United Airlines/California Airline Pilots Assoc
Coleman	**Bill**	Chairman	BEA Systems, Inc.
Jesse	Frank	VP Real Estate & Corp. Services	BEA Systems, Inc.
Crowell	**Bill**	President & CEO	Cylink Corporation
Cuéllar	**Mariano-Florentino**	Assistant Professor	Stanford University, School of Law
England	**Sandra**	Exec. Vice President	Network Associates
Blough	Kelly	VP, Investor, Gov't. & Community Relations	Network Associates
Sabo	Doug	Manager, Govt & Community Relations	Network Associates
Fox, Sr.	**Michael E.**	Chairman	M.E. Fox & Company
Harris	**Don**	Director of Special Projects	Southwest Airlines
Hearney	**Richard**	General (ret.) President & CEO	BENS
Infante	**Beatriz V.**	Chairman, President & CEO	Aspect Communications
Morgante	Guy T.	VP, Vertical Solutions	Aspect Communications
Lansdowne	**Bill**	Chief of Police	San José Police

Appendix A – Blue Ribbon Task Force Participants

Last Name	First Name	Position	Organization
Bold = Task Force Members			
Lewis	Steven L.	Lieutenant	San José Police, Airport Division.
Magistri	**Dr. Sergio**	President & CEO	InVision Technologies, Inc.
McCashin	**Bob**	Chairman & CEO	Identix
Scullion	Jim	President & COO	Identix
Nishimura	**Dr. Ko**	Chairman & CEO	Solectron Corporation
Fok	Phil	CAO	Solectron Corporation
Palmer, Jr.	**Richard W.**	Vice President	Cisco Systems
Panu	**Krish**	CEO	@Road
Wansley	**Larry A.**	Managing Dir. of Corp. Security	American Airlines
Franco	Tedeschi	General Manager, San Jose	American Airlines
Weidemeyer	**Tom**	COO-UPS, President, UPS Air	United Parcel Service
Weigle	**Peggy**	President & CEO	Sanctum
Advisors to the Task Force			
Withycombe	Bill	Regional Administrator	FAA Western Pacific Region
Anthony	Thomas	Regional Director	Transportation Security Administration (TSA)
Office of City Manager, San José			
Dearborn	Darrell	Senior Deputy City Manager	City of San José
Mineta San José International Airport			
Tonseth	Ralph G.	Director of Aviation	Mineta San José International Airport
Kunesh	Cindy	Admin. Assistant	Mineta San José International Airport
Peterson	Jim	Deputy Director	Mineta San José International Airport
Reinhardt	Debbie	Secretary	Mineta San José International Airport
Felix	Charlie	Information Systems Manager	Mineta San José International Airport
Luckenbach	Steve	Communications Manager	Mineta San José International Airport
Smith	Jamie	Senior Analyst	Mineta San José International Airport
Consultants			
Gerston	Larry N., Ph.D.	Principal Writer	Gerston & Associates
Gregory-Grant	Callie	Project Coordination	Consultant
Claggett	Sunny	Task Force Facilitator	Shilling & Kenyon

APPENDIX B

Committee Chairs, Members and Staff

B-1

Appendix B --Committee Chairs, Members and Staff

Technology Demonstration Committee

Committee Chair

Beatriz V. Infante — Chairman, President & CEO — Aspect Communications

Name	Title	Company
Thomas Anthony	Regional Director	Transportation Security Administration
M. Sam Araki	Chairman	Security Technology Ventures, LLC
Captain Dan B. Ashby	Chairman, U. A. Council 34-SFO	California Air Line Pilots Association
Matthew Bostick	Field Representative	Office of Congressman Mike Honda
Sandra England	EVP Bus Dev & Strat Research	Network Associates
Brian Finan	Director, Symantec Federal	Symantec Corporation
Michael E. Fox, Sr.	Chairman	M.E. Fox & Company
Don Harris	Director of Special Projects	Southwest Airlines Co.
Frank Jesse	VP Real Estate & Corp Svcs	BEA Systems Inc.
Lt. Steven L. Lewis	Operations Commander	San Jos Police, Airport Division
Deborah Magid	Director, Strategic Alliances	IBM
Jim Peterson	Deputy Director	Mineta San Jose International Airport
Richard W. Palmer, Jr.	VP & GM VPN & Security Svcs	Cisco Systems
Krish Panu	Chairman & CEO	@Road
Jim Scott	President	Recognition Systems
Jim Scullion	President & COO	Identix
Ralph G. Tonseth	Director of Aviation	Mineta San Jose International Airport
Peggy Weigle	CEO	Sanctum, Inc.
Bill Withycombe	Regional Administrator	FAA Western Pacific Region

Technology Demonstration Committee Staff		
Tim Duffy		Arcsight
Rod Fan	Chief Technology Officer	@Road
Larry Sells		Aspect (Gov)
Bill Follin		Aspect (Gov)
Robert Caruso		Recognition Systems
Paul Haight		Cisco Systems
Martin Huddart	General Manager	Recognition Systems
Gareth Owen	Chief Technology Officer	Aspect Communications
Steve Orr	Global Airline Segment Executive	IBM
Gary Barnett	EVP, CTO	Aspect Communications
Guy Morgante	VP, Vertical Solutions Dev.	Aspect Communications

Proposal Development Committee

Committee Chair

Michael E. Fox, Sr. — Chairman — M.E. Fox & Company Inc.

Name	Title	Company
Tino Cuellar	Assistant Professor, School of Law	Stanford University
Gen. Richard Hearney	President & CEO	BENS
Richard W. Palmer Jr.	Vice President, VSEC Business Unit	Cisco Systems
Bill Crowell	President & CEO	Cylink Corporation
Sam Araki	Chairman	Security Technology Ventures

APPENDIX C

Task Force Meeting Summaries

Appendix C — Task Force Meeting Summaries

Kick-off Meeting, March 18, 2002

TASK FORCE ATTENDEES

Congressman Mike Honda, Honorary Co-Chair
Mayor Ron Gonzales, Honorary Co-Chair
John W. Thompson, Symantec, Task Force Chairman

Members:
Sam Araki, Security Technology Ventures
Capt. Dan Ashby, CA Airline Pilots Association
Tino Cu llar, Stanford University Law School
Bill Crowell, Cylink
Sandra England, Network Associates
Mike Fox, Sr., M.E. Fox Company
Don Harris, Southwest Airlines
Gen. Richard Hearney, BENS
Beatriz V. Infante, Aspect Communications
Bob McCashin, Identix
Chief Bill Lansdowne, San Jose Police Department
Sergio Magistiri, InVision
Richard Palmer, Cisco Systems
Krish Panu, @Road
Tom Weidemeyer, UPS
Peggy Wiegle, Sanctum
Bill Withycombe, FAA Western Pacific Region

Liaisons:
Dan Perez for Ko Nishimura, Solectron
Franco Tedeschi for Larry Wansley, American Airlines
Jim Scullion accompanied Bob McCashin, Identix

AGENDA

Welcome & Introductions:

- Task Force Chairman John Thompson welcomed the Task Force members and their liaisons to officially launch the 100-day period for the group s work. In addition to introducing the members, Chairman Thompson emphasized his commitment to making travel through the San Jose airport more predictable and consistent while also meeting security requirements.

Task Force Objectives:

- Congressman Honda thanked the group for signing on to the Task Force and encouraged the free exchange information to create a program that will maintain a commitment to privacy and personal freedom. He said the nation will appreciate the Task Force s work and is anticipating the results.

Appendix C — Task Force Meeting Summaries

- As a member of the House Transportation Committee s Aviation Subcommittee, Congressman Honda anticipates an update on the application process for Federal technology pilot programs in the coming weeks. The Task Force s final report will make the San Jose airport a contender for pilot status and the associated Federal funding.

- Mayor Ron Gonzales gave his appreciation to the chairman and Task Force members on behalf of the City of San Jose. He said that the group, many of whom are frequent travelers, will gain a better understanding of the challenges ahead, making it important to focus some of the best minds in the Valley on meeting the new FAA regulations, as well as anticipating future requirements.

- Mayor Gonzales said the Task Force is to identify and examine technologies in passenger security, personnel, baggage screening (e.g. bomb detection) and airfield security that will also improve the customer experience. He stressed the 100-day timeframe for the Task Force s work, which will include gathering public input.

- Once the recommendations are assembled, they will be presented to the San Jose City Council and the U.S. Department of Transportation with the goal to become a pilot program.

Meeting and Decision-Making Process:

- Chairman Thompson requested volunteers to chair two subcommittees: technology and proposal development. Beatriz V. Infante offered to chair the technology subcommittee and Mike Fox, Sr. volunteered to chair the proposal development subcommittee. The subcommittee chairs will recruit members for their two areas.

- Chairman Thompson discussed the contents of a background binder presented to members at the meeting (binders are to be sent to those who teleconferenced for the meeting). He described the meeting process, which will include 5 regular meetings for various industry and government panels, one public hearing, and a wrap-up session for final adoption.

- Chairman Thompson concluded this portion of the meeting with the anticipation that the Task Force could recommend a minimum of six technology applications that will meet the stated goals. He stressed that the group will remain focused on technology and how it impacts customer service, and that it is not charged with addressing other issues such as parking, personnel or buildings.

Questions and Comments:

- Bill Crowell of Cylink wanted to ensure the group would not overlook process architecture in considering technologies. Chairman Thompson concurred that the final proposal will have to incorporate process change in its recommendations.

- Beatriz V. Infante added that technology should be maximized because in relying on people to handle a process there is potential fallibility that part of the uses of technology will be in automation to help eliminate error.

Appendix C — Task Force Meeting Summaries

- Mike Fox asked whether general aviation security would be included as part of the proposal, and Chairman Thompson affirmed that representatives in one of the Task Force meetings will address the issue.

- Bill Withycombe requested that binders be mailed to members calling in.

- Beatriz V. Infante asked about the appropriate size for subcommittees. Chairman Thompson said the best measure would be how many are required to do the job, and he gave the subcommittee chairs discretion.

- Tino Cu llar pointed out the need to understand the different security threats that the group must address. He asked whether the final proposal must outline costs for the recommended technology. Chairman Thompson agreed that the Task Force should be briefed on security threats. He also said a certain level of pragmatism will have to be employed in the process so that practical, workable solutions are recommended.

- Symantec Public Affairs Manager Adam Rak announced the development of a password-protected Web site under development for the administration of the Task Force s work. Details will be provided as soon as they are finalized.

Chairman Thompson adjourned the meeting. The first formal Task Force meeting* is scheduled for Friday, April 5, from 1 p.m. to 4 p.m. in San Jose at the Mineta International Airport administration offices, 1732 N. First Street. The meeting topic will be The Federal Government View and Airline Industry Challenges.

###

***Corrections and amendments will be incorporated at the next Task Force meeting.**

Appendix C — Task Force Meeting Summaries

Minutes of April 5, 2002
BLUE RIBBON TASK FORCE

**For a tape recording of the meeting, please contact Cindy Kunesh at 408-501-7669*

1. **Welcome by Chairman John W. Thompson**

2. **Attendance**

Chairman John W. Thompson, Symantec
Sam Araki, Security Technology Ventures
Capt. Dan Ashby, Airline Pilots Association/United Airlines
William Crowell, Cylink Corporation
Sandra England, Network Associates
Mike Fox, Sr., M.E. Fox & Company
Don Harris, Southwest Airlines
Beatriz V. Infante, Aspect Communications
Chief Bill Lansdowne, San Jose Police Department
Sergio Magistri, InVision Technologies, Inc.
Krish Panu, @Road
Tom Weidemeyer, UPS
Peggy Weigle, Sanctum
Bill Withycombe, FAA Western Pacific Region

Other:
Ralph Tonseth
Phil Fok
Frank Jesse
Bill Scullion
Paul Haight
Meri Maben
Jim Webb
Adam Rak
Cris Paden
Jim Peterson
Sunny Claggett
Callie Grant
Doug Sabo
Lt. Steven Lewis
Charlie Felix
Leslee Coleman

Appendix C — Task Force Meeting Summaries

3. **Announcements**

Chair Thompson announced that there have been a number of press inquiries about the Task Force s effort and reminded the members to either funnel all inquiries to Congressman Honda s Office or the Mayor s Office, in order to communicate a consistent message to the public. The contact in the Congressman s Office is Ernest Baynard and David Vossbrink for the Mayor s Office. All suggestions or inquiries about technologies should be forwarded to Beatriz V. Infante, Chairman of the Technology Subcommittee.

The next Task Force meeting will be held on April 17th at Airport Headquarters, from 1 — 4 pm.

Due to time constraints, demonstrations for technology will need to be shifted around. The calendar will change, but it will give the committee more time to review many of the technologies presented. The public hearing will now be May 10 and the technology review meeting will May 31.

4. **Subcommittee Reports**

Deferred until after the presentations.

5. **Presentations**

a. Robin Hunt — Program Director for Aviation Security and Infrastructure, U.S. Department of Transportation, OIG.

Ms. Hunt reported that the Department of Transportation has been doing audits and investigations of aviation security since the early 1990 s. Numerous reports have been issued over the years, including audits and testimonies on access control, cargo security, and deployment of advanced technology.

Three separate audits have recently been initiated. In January, an audit was begun on advanced security technology. We are looking at the types of technology that are out there, either ready for deployment or long term, but show promise, with additional funding, to hopefully being deployed in the airport environment.

A second audit, which was just begun on Tuesday, April 2, 2002, is looking at TSA s hiring process and deployment of screeners. The focus will be on selection, training, and a plan to employ those screeners.

Appendix C — Task Force Meeting Summaries

A third audit, also begun this week, is focused on the deployment of advanced technology. While we re looking at all technology being deployed; the focus is really on explosive detection systems.

One of the challenges TSA faces is hiring and training screeners by November 19th, 2002. As many as 60,000 may need to be hired. At the larger airports, as many as 80% of current screeners may not qualify because they are not U.S. citizens.

Another challenge TSA faces is the deployment of the explosive detection systems; the deadline is December 31, 2002. Many airports across the country will have to be reconfigured at an enormous expense.

The final challenge is budget. The estimated cost is between $4 and $6 billion. And it has not been decided who will actually fund the installation of these systems.

b. Thomas Anthony — Manager of the Civil Aviation Security Division of the Western Pacific Region, TSA

Mr. Anthony began by stating that the most obvious security change is the fact that the Federal Government is now responsible for passenger screening. All airports that provide commercial passenger service, whether small or large, will be affected by this change. It is important to keep in mind that there are different applications of security technology for different airports. TSA needs to establish control and a level of service at security checkpoints that will be acceptable to the traveling public.

General Aviation is another area of concern and TSA is looking into the security screening of these travelers and their aircraft. The crew and passengers will be screened prior to boarding the aircraft if greater than 12,500 pounds. Also, any non-U.S. citizen who applies for instruction in any type of aircraft must now make application to the Department of Justice. Only with the DOJ s approval may they take flight lessons.

c. Don Harris, Director of Systems Projects — Ground Operations, Southwest Airlines.

Mr. Harris reported that Southwest Airlines operates approximately 2,920 daily departures out of 59 domestic airports. Security checkpoint queue lines, in many cases, are the primary frustration for travelers.

By far, Mr. Harris said, the largest challenge is the Computer Assisted Passenger Profiling System (CAPPS). Because Southwest Airlines does not issue boarding passes, passengers queue up to board the plane in sections, with section one boarding first. If you are queued up to board first and become a CAPPS selectee, you lose your place in the boarding process.

A CAPPS selectee cannot be processed at curbside check-in. The ticket counter must designate a position to process Selectee Customers, thereby limiting the number of positions that are available to process remaining customers. Selectee customer processing at the departure gate is the primary source of complaints from customers.

Mr. Harris went on to say that Passenger/Bag Match (PBM) requirements slow the activity of loading and unloading aircraft. New directives come about often and quickly and arrive with little warning and limited time to prepare for operation compliance. This creates confusion for personnel and customers.

Many new technology and automation projects, intended to increase passenger processing capacity and enhance our ability to comply with security mandates, are being considered at an accelerated pace. Our challenge is to maintain operation simplicity without dramatically altering customer satisfaction.

d. Thomas Weidemeyer — Chief Operating Officer of UPS and President, UPS Airlines

Mr. Weidemeyer stated that the primary job of UPS is to keep commerce moving. UPS is the 11th largest airline in the world, and 90% of its customers are regular customers and do business on a daily basis. The balance between security, convenience, efficiency and passenger experience is vital to the future health of the industry. The cargo industry is unique, however, because one size does not fit all.

He stressed that information technology is very important to the business of UPS, in the detection of radioactive, chemical, and biological hazards. Moreover, knowing their customers so intimately takes the anonymity out of the equation. He stressed convenience as an important business factor.

Chair Thompson asked Mr. Weidemeyer what was the most significant impact on UPS since September 11th.

Mr. Weidemeyer replied that the impact was not in the aviation arena, but in the ability to serve customers.

Appendix C — Task Force Meeting Summaries

6. Question and Answer Period Highlights

Mr. Crowell asked if the committee could know, in a general sense, if the CAPPS program results in real incidents, that is a serious security threat, and to what extent does the airline view this program as value added to the security program. Also, has Southwest given any thought to moving this process up earlier into the arrival at the airport. Will the regulatory environment allow Southwest to move the process up?

Mr. Harris replied that there is very little that can be done to prepare for the selectee s arrival other than the ability to increase the security capacity.

Mr. Crowell asked if a passenger was checking into the main gate at the airline and was tagged for secondary screening, why wouldn t you do the security check then, as opposed to waiting for them to get through the first security barrier and then screening them at the gate?

Mr. Harris replied that, indeed, the security checkpoint is the most painful bottleneck in the process.

Capt. Dan Ashby stated that most of the airlines don t have their own terminals. In order to look at the selectee early on, you will move the gridlock further into the street. Someone that may trigger CAPPS over in Southwest may be different from some sort of other criteria another airline may use. Captain Ashby added that screeners currently working at checkpoints are not trained to make judgments after looking at the facts.

Mr. Crowell replied that he was trying to ask the question, Who is responsible for security? Is it the airlines or TSA? If the airlines are going to retain responsibility for checking people a second time, then this is a dilemma that could be very difficult to solve process wise.

Chair Thompson stated that it s one thing to focus on the process, and security is both about technology and process.

(A lengthy discussion followed regarding bar coding on e-tickets, faxed confirmations, and a national database.)

Chair Thompson called for any further questions or comments.

Appendix C — Task Force Meeting Summaries

Mr. Withycombe, FAA, Western Pacific Region, commented on the fact that the FAA has a great responsibility for national airspace system and moving aircraft through the system on a record basis. Keeping the delays to a minimum is always a major project on a day-to-day basis. Not only fighting the weather elements, but also the fact that when the terminal is dumped (as the result of a perceived security threat), it creates havoc throughout the system. This affects not only the immediate terminal area and most flights departing, but it has a rolling effect that goes all the way across the country. Another area that he thought should be addressed very carefully is checked baggage and increasing efficiency in the system while also applying this technology.

Mr. Withycombe added that Thomas Anthony s area of expertise is aviation security and suggested that Tom would be an excellent resource for the committee. With that recommendation, Chair Thompson welcomed Mr. Anthony as an official member of the Blue Ribbon Task Force.

4. Subcommittee Reports

a. Beatriz V. Infante, Chair of the Technology Subcommittee reported that, at the request of the Mayor s Office, they have surveyed a process for solicitation of emerging technologies. Ms. Infante stated that she has received numerous creative suggestions via e-mail, most of which would not be technologically feasible. In order to enforce some structure on this project, people will be required to fill out a template of a Request for Information (RFI) or Request for Proposal (RFP). This template will be online and will enable the Technology Subcommittee to evaluate the proposals in a more uniform manner. It is expected that this template, which is being developed with the guidance of Gartner Group, should be on the website within the next week.

   At this point, Chair Thompson asked Adam Rak, Government Relations Manager, Symantec, to comment on the website. Mr. Rak, presented a PowerPoint slide of what the Internet home page will look like. Certain areas of the website will be password protected for Task Force Member access only.

b. Mike Fox, Sr., Chairman of the Proposal Development Committee, reported that they are waiting for proposals to come in. Two members of the committee are Tino Cuellar and Richard Hearney. Richard Palmer may also join the committee. Mr. Fox suggested having some members of the Technology Subcommittee serve on the Proposal Development Subcommittee since they are going to be the genesis of the content; Ms. Infante agreed.

Chair Thompson then encouraged each member of the Task Force to join a committee and become actively involved. He also reminded the Task Force members that they have just 100 days to research and complete a report.

There being no further questions or comments, Chair Thompson adjourned the meeting at 3:25 pm.

Appendix C — Task Force Meeting Summaries

Minutes of April 17, 2002
BLUE RIBBON TASK FORCE

**For a tape recording of the meeting, please contact Cindy Kunesh at 408-501-7669*

1. **Welcome by Chairman John W. Thompson**

2. **Attendance**

Chairman John W. Thompson, Symantec
Thomas Anthony, TSA, Task Force Advisor
Sam Araki, Security Technology Ventures
Capt. Dan Ashby, Airline Pilots Association/United Airlines
Bill Coleman, BEA Systems
William Crowell, Cylink Corporation
Sandra England, Network Associates
Mike Fox, Sr., M.E. Fox & Company
Don Harris, Southwest Airlines
Beatriz V. Infante, Aspect Communications
Chief Bill Lansdowne, San Jose Police Department
Krish Panu, @Road
Bob Bergman for Tom Weidemeyer, UPS
Peggy Weigle, Sanctum

Other:
Chip Barclay
Kelly Blough
Ralph Tonseth
Frank Jesse
Lt. Steve Lewis
Doug Jones
John Aitken
Matthew Bostick
Meri Maben
Jim Peterson
Callie Grant
Chris Paden
Charlie Felix
Jim Webb
Sunny Claggett

3. ANNOUNCEMENTS

Chris Paden, Public Relations Manager, Symantec, reported that the Blue Ribbon Task Force (BRTF) Web site was up and running as of Monday, April 15th. The general public will be able to access the site. A special section of the Web site, accessible by password, will be for BRTF members only. The Web site will be the main receptacle for applications to be considered by the Technology Subcommittee. Press releases will also be posted on the site. Mr. Paden distributed an update of media releases to date as well as those planned for future release. He also reminded the Task Force that if the press contacts them, to please contact the Congressman s Office, the Mayor s Office or Chris, himself.

Beatriz V. Infante, CEO, Aspect Communications, said that it is important to figure out the right timing for closing the RFI process and having the public hearing. Following a discussion, the group agreed to extend the RFI deadline to May 10.

Chair Thompson addressed the issue of protecting the disclosure of propriety security processes important to the work of the Task Force. He recommended that a Non-Disclosure Agreement (NDA) would be an appropriate document for all members to sign. In that way, participants in the Task Force would feel free to share information. Chris Paden will be faxing an NDA to each Task Force member to be signed before the next meeting.

3. SUBCOMMITTEE REPORTS

a. Beatriz V. Infante, CEO, Aspect Communications, reported that the first meeting of the Technology Subcommittee would be Friday, April 19th. Many members of the Task Force have volunteered to serve on the committee.

b. Mike Fox Sr., President, M.E. Fox and Company, reported that the Report Writing Committee had their first meeting on April 17th. The committee has three members to date: William Crowell, President & CEO, Cylink Corporation, Professor Tino Cuellar, Stanford University School of Law, and Richard Hearney, President & CEO, BENS.

Appendix C — Task Force Meeting Summaries

4. **PRESENTATIONS**

a. Charles Barclay, President, American Association of Airport Executives (AAAE), presented the organization s operational roles with respect to security. AAAE, in partnership with the Federal Government, runs the Aviation Security Clearinghouse, which includes coordinating fingerprinting and recording keeping certain personnel for 429 airports and dozens of airlines. AAAE also provides computer-based training for security at key airports, such as for Salt Lake City during the Olympics. AAAE creates custom programs to analyze airport security and train personnel. It is a turnkey solution that can be put in place for any airport s specific system. San Jos has recently ordered this type security training.

AAAE also maintains a satellite base for a distance learning and training network that was begun under the leadership of Ralph Tonseth, a past chairman of AAAE and director of the Mineta San Jos International Airport. It is mostly a training network, but it also allows for public affairs programming. For instance, Secretary Mineta has done a call-in show. It has been a very valuable tool for getting information to airports.

In reference to Sept. 11th, Mr. Barclay said the problem now is not the physical threat to the system, but the economic threat. The inefficiency of the system is so great, that airlines can t make a profit. Today, American Airlines is still losing $4,000,000 a day, and the other network carriers continue to bleed cash. Finding a balance of safety, security, convenience and customer service is critical.

TSA is trying to come up with a single transport worker identification card that will be universal. The problem airports are having is that GSA standard and the Smart Card do not have a biometric on it now. Since that is a government process, it will take some time to add a biometric to it. The airports are under pressure to add biometrics to current access control systems, but if that is done on an interim basis it will have to be thrown out in the future if it is not compatible with what TSA is intending to do.

The biggest hurdle for airports today is the mandated EDS installation by December 31st. As of yet, most airports don t even know the basics what the mix will be of these extremely large and very expensive machines. AAAE has conveyed in Washington that EDS can be built correctly into the system, but it is an impossibility to do all that by December 31, 2002; an interim solution will be necessary.

MR. BARCLAY SUMMED UP HIS PRESENTATION BY SAYING THAT DEVELOPING NEW TOOLS FOR NETWORKING INFORMATION IS VITAL. THE THREAT TO THE SYSTEM, THE ONE THING THAT IS BRINGING THE SYSTEM TO ITS KNEES, IS TOO FEW PASSENGERS. CONVENIENCE, CUSTOMER SERVICE AND EFFICIENCY MUST BE BUILT BACK INTO THE SYSTEM, AS WELL AS SECURITY.

b. John Costas, Deputy Airport Director, Chief of Staff, San Francisco International Airport, began his report by saying that the government alone, TSA, is not going to solve the problem. The challenge requires the partnership and the collaboration of TSA, government, airlines, airports, and technology providers. The airport industry has many challenges in developing and applying technology. The focus needs to be on research and development in the area of high threat assessment needs, such as federal certification, availability and production capability, reliability and maintainability, standardization vs. customization, mobility, obsolescence, and acquisition / O&M cost.

Funding is another challenge for the industry. The airports remain un-reimbursed for over $325 million in post 9/11 security expenses. Many airport infrastructure projects are in competition with security requirements for AIP funds. Since 9/11, significant airport revenue losses have accrued; SFO alone sustained a $100 million loss.

TSA mandates pursuant to the Aviation Security Act are not fully scoped and what is currently identified does not match existing funding levels. TSA has budgeted $175 million for each EDS unit to be installed; this is inadequate for in-line systems. Another $4.4 billion has been requested by TSA primarily for personnel expenses and only 1 billion for explosive detection equipment. TSA is not budgeting for nor contemplating the loss of airport revenue space to conduct TSA security and support operations. Much ambiguity still exists between the TSA, airports and airlines regarding responsibility of security costs.

Mr. Costas summed up his report by expressing the concerns SFO is facing. Among them is the extremely low probability of satisfying full EDS mandate by December 2002. The in-line installation is the only feasible method without causing severe congestion and affecting customer service. The reliability and maintainability of EDS units is historically wrought with problems.

c. Ralph Tonseth, Director of Aviation, Norman Y. Mineta San Jos International Airport, said that in his view a main theme of this task force was to have the passengers feel confident that they are safe, and that it is convenient to do so; the process of air travel needs to be as non-intrusive as possible. The major issues facing SJC today include screening improvements, securing perimeter improvements, terrorist protection and mitigations measures, and maintaining high levels of expertise and responsiveness.

   Air passenger behavior has changed dramatically since 9/11. For example, security-screening procedures take longer; lines at security checkpoints can be extremely long during peak hours; passengers arrive earlier and stay longer; and meeters and greeters are not permitted past security.

   Mr. Tonseth said that terminal security responses, secure area improvements, belly cargo screening, air cargo security and screening, parking facility measures, and federal agency space are the six projects SJC has identified in order to address security measures.

   In summing up his report, Mr. Tonseth, said that technology is being called upon to increase productivity, accuracy and dependability to address the national challenges facing the aviation industry. Industry airline expertise is necessary to restore public assurance about airport convenience and aviation safety.

d. Laura Simpson, Manager of Customer Service, Southwest Airlines, SJC gave a report on how security measures have affected customer service. Security screening at the gates. What used to be a very quick process is now being impacted in that passengers are being asked to take off their shoes, have their bags searched, show personal identification and boarding documentation; all this has to be done in a very expeditious manner in order to get people on the plane. Time spent with a customer one-on-one is cut considerably. The message our bag checkers and airline employees are receiving is that the overall experience passengers are receiving is tedious and intrusive. Ms. Simpson suggested that monitors showing a video to passengers while they re waiting in line of what s expected of them would be very advantageous. But it is also necessary to be aware of information overload. On a busy day, Southwest Airlines will process as many as 9,000 passengers. Prior to 9/11 passengers weren t required to arrive at the airport at 4:00 or 4:30 am; now they are doing so in order to board their flight on time. Ms. Simpson summed up her report by saying she felt that with excellent technology and good personnel, the security screening process could be improved.

e. Marc Casto, Vice President of eCommerce and Fulfillment Services, Casto Travel, reported that Casto Travel is one of the larger travel agencies in Silicon Valley, and in the top 30 of the United States. Even though Casto Travel had emergency plans for an earthquake or fire or power outage, there were no plans for a terrorist attack that would ultimately shut down travel agencies all across the United States for a couple weeks. On September 11, 2002, Casto Travel received two type of phone calls from clients: 1) Were any of my employees on those planes? 2) Where are my employees right now? In order to get that information, they had to invent a new operating system within the current database to pull the much-needed information. It took about two and a half hours to assemble the information, which was then distributed to corporate clients, so they in turn could notify family and other colleagues of the travelers status. Following that action, Casto Travel had to contact travelers with advice of alternate travel. There were about 10,000 people ticketed by Casto Travel who were at different places throughout the world on 9/11. With a staff of 250 people, it was overwhelming to contact 10,000 people in such a minimal amount of time, each traveler with a different issue and how to get home. Rental cars, Amtrak, Greyhound, motor homes, etc. were all utilized in order to get home.

THE STATUS OF THE TRAVEL INDUSTRY, GIVEN 9/11, IS WAY DOWN; A LARGE NUMBER OF AGENCIES ARE CLOSING THEIR DOORS. OVER 14% OF AGENCIES IN THE U.S. FROM MARCH 2000 TO FEBRUARY 2002 HAVE CLOSED THEIR DOORS. ANOTHER SIGNIFICANT FACTOR IS THAT THE AIRLINES USED TO PAY COMMISSION TO MANY TRAVEL AGENCIES, A STRONG FORM OF REVENUE FOR THE TRAVEL AGENCIES. THIS IS NO LONGER THE CASE AND MANY OF THE SMALLER COMPANIES ARE NOT GOING TO BE ABLE TO SURVIVE IN THIS NEW ECONOMIC ENVIRONMENT.

Mr. Casto concluded by saying travel agencies are still the distribution point for about 75 — 80 % of all airline tickets issued. The other 10 to 15% are purchased through the airlines themselves or on line.

5. FACILITATED DISCUSSION OF TASK FORCE GOALS AND OUTCOMES

Highlights

Chairman Thompson introduced Sunny Claggett, a consultant with SK Consulting, to help the group synthesize the presentations and the members reactions. Ms. Claggett suggested four common themes to keep in mind:

i. This is a period of great change in airline/airport security.
ii. A large challenge is to balance time, security, cost, and convenience.
iii. Recommendations must be a combination of technology and people.
iv. Improving security is an evolving process.

During the next 30 minutes, an interactive discussion took place identifying the issues that are foremost. Following are some of the highlights:

William Crowell began with the question, Who s in charge?
There is an inherent conflict about who is in charge. While TSA thinks they are in charge, as is mandated by the law, they don t control the purse strings for the potential implementation that will have to occur airport by airports around the country. And so how can you in fact be in charge and mandate solutions without being able to offer up the financial capability to implement this?

Beatriz V. Infante said she felt part of the challenge is that who s in charge varies month to month, so there is a whole migration of what are traditionally local or state responsibilities being taken over by a centralized federal responsibility.

Capt. Dan Ashby said that leadership is of the utmost importance and then funding. But before you develop the hierarchy, the premise of zero tolerance must be in place. The commitment has to be there to follow through with the mission or it s all just dialogue.

Tino Cuellar stated that to understand exactly what control TSA has over the process is vital. On the zero tolerance issue, the problem is defining what we mean by zero tolerance not an easy thing to do and it s not self-explanatory. It s possible to have a security system that it so perfect that absolutely almost any risk would be reduced, but then who would fly? So it needs to be clarified what sort of trade offs we re expecting passengers to make about the information they are providing.

Sam Araki said perhaps the way to go about it is to implement preventive security

Peggy Wiegle said that what TSA has proposed can t be implemented and so no one is going to sign up for the objectives. Fundamentally, she said, we are at a very difficult place and looking at it from a different angle is a better way to go about it.

Beatriz V. Infante said the focus should be on technologies that are more data intensive, more upfront, more preventive, going all the way back to the airline reservation process.

William Crowell agreed with Beatriz, that the focus of security should begin early in the process; access control is only a small piece of the process. Some decisions have to be made about how to address the basic conflict of what the law says and what this task force is trying to accomplish.

Chip Barclay followed up by saying that there is never going to be enough money to do all of this, so it s best to invest the technologies where they can have the most impact by looking at the points where you can, in fact, have the most influence.

Beatriz V. Infante said that one of the things the committee may want to make a recommendation on is the notion of trusted traveler — is it a good thing or a bad thing?

William Crowell stated that having a badge is not going to be a certainty that someone is not going to do something. The system has to be multi-layered.

Chairman Thompson reminded the committee that SJC is one of the largest general aviation airports in the country, so that portion of the traveling public needs to be taken into consideration. Also, air cargo is a huge issue as well. The focus should not be concentrated on just passengers alone.

Sandra England stated that understanding the process is most vital and the committee does have to map this out from the point of purchasing a ticket to actually getting on the airplane. In the security industry it is important to focus on vulnerability assessment, risk assessment, understanding the tradeoffs — these are all things that need to be put in place, but first the process must be understood.

Krish Panu mentioned that one other dimension to be added to cost, security and convenience is privacy. It is a factor that will certainly come up as a trade off.

Chair Thompson concluded that the brainstorming session certainly helped to narrow down the task at hand.

5. APPROVAL OF MINUTES

Chair Thompson asked if there were any additions, deletions or comments to the April 5 meeting minutes. There being none, he called for a motion to approve the minutes. Chief Lansdowne moved that the minutes be approved and Tino Cuellar seconded the motion. The motion passed unanimously.

6. ADJOURNMENT

Before the meeting adjourned, Chair Thompson reminded the task force that the next meeting will be a public hearing held on May 10th at the Silicon Valley Conference Center located at 2161 N. First Street. The task force will meet at the Conference Center from 1 p.m. to 1:30 and then participate in the public hearing from 1:30 pm to 4 pm. A call-in line for members will be available for both segments.

Meeting adjourned at 4:10 pm.

Appendix C — Task Force Meeting Summaries

Minutes of May 10, 2002
BLUE RIBBON TASK FORCE
Silicon Valley Conference Center
2161 North First Street
San Jos , CA

Taskforce Meeting: 1:00-1:30 pm
Public Hearing: 1:30-4:00 pm

**For a tape recording of the meeting, please contact Cindy Kunesh at 408-501-7669*

I. Welcome

II. Roll Call

Chairman John W. Thompson, Symantec
Sam Araki, Security Technology Ventures
Tino Cu llar, Stanford University
Sandra England, Network Associates
Mike Fox, Sr., M.E. Fox & Company
Don Harris, Southwest Airlines
Gen. Richard Hearney, BENS
Sergio Magistri, InVision Technologies
Richard Palmer, Cisco Systems
Krish Panu, @Road
Bill Withycombe, FAA
<u>Other:</u>
Frank Jesse
Pat Reilly
Jim Scullion (public hearing)
Phil Fok
Grant Evans
Guy Morgante
Ralph Tonseth
Meri Maben
Jim Webb (public hearing)
Cris Paden (public hearing)
Brian Finan
Sunny Claggett
Callie Grant
Cindy Kunesh
Lt. Steven Lewis
Charlie Felix
Matthew Bostick
Sue Knill

Appendix C — Task Force Meeting Summaries

III. Review and Approval of Minutes from 4/17 Meeting

Chair Thompson called for a motion to approve the minutes of April 17th. A motion was made and seconded. Motion carried unanimously.

IV. Fieldwork Opportunity

Ralph Tonseth, Director of Aviation at Mineta San Jose International Airport, invited the Task Force members to serve as Airport Ambassadors for about an hour on May 24th in Terminal A. This would be a good opportunity for committee members to experience first hand the concerns and issues of the general public.

V. Subcommittee Reports

Technology Committee: Guy Morgante, Aspect Communications, distributed an outline for the technology demonstration process. PowerPoint and hands-on demonstrations are encouraged, not more than 15 minutes in length, for the meeting on May 31, 2002.

At the Task Force s June 4 meeting, the subcommittee will make recommendations based on: 1) Technology Landscape; 2) Passenger & Workforce Process Flows; 3) Critical Risk Areas; 4) Technology applications to Passenger & Workforce Critical Risk Areas.

Report Writing Committee: Mike Fox, Sr., M.E. Fox and Co., reported that subcommittee members have begun working on the outline and the integration of proposals into what will become the final report.

Mr. Fox introduced Larry Gerston, Professor of Political Science at San Jos State University, who will assist with the development of the report.

VI. Public Hearing Briefing

Sunny Claggett stated that the invited presenters would have 5 minutes to present their product, idea, or concern, and the general public 2 minutes. The media will most likely be in attendance as well.

VII. Adjournment

Meeting adjourned at 1:30 pm.

Appendix C — Task Force Meeting Summaries

BLUE RIBBON TASK FORCE

Public Hearing
May 10, 2002
1:30 pm

Silicon Valley Conference Center
2161 North First Street
San Jos , CA

Chair Thompson opened the public hearing by stating that the mission of the Task Force was to develop and submit a report to the Transportation Security Administration in June, outlining recommendations of technologies that can improve security and the security process at our nation s airports. It is also important, he noted, to have the traveler experience improved as well. In light of 9/11, the focus of the Task Force s effort is on three fundamental attributes: cost, convenience and security. The critical element of this mission is to cast the broadest net possible in gathering information and knowledge of the challenges facing aviation security today. Providing the opportunity to hear what the public has to say about these issues will play a key role in ensuring that the efforts are thorough and thoughtful, he said.

Mr. Thompson announced that it is assumed by the Task Force that any information that a presenter submitted here today, is not proprietary information and that it can become a part of the public record. He also said that if someone had an interest in submitting a RFI based upon the submission deadlines that have been set, that RFI submission should be in by midnight, May 10' 2002.

Following is a complete listing of the presenters with a brief note on the subject of their testimony:

1. Steve Kirsch demonstrated new iris scanning technology.
2. Richard L. Rowe, self-employed, presented his ideas on biological sensor technology.
3. William H. Dunlop, Lawrence Livermore National Lab, advised on radioactive sensor technology, which is currently being deployed in Russia.
4. Katie E. Corrigan, American Civil Liberties Union, Legislative Council on Privacy Issues, spoke on privacy, equality and fairness.
5. Rod Dewell, Excalibur Solutions, Inc., a private pilot and biometric technologist, spoke on applying biometric technology in the cockpit.
6. Michelle Kraus, Ping I.D., spoke for digital identity.
7. Helal R. Omeira, Council on American-Islamic Relations, also spoke of privacy, equality and fairness.
8. Jim Cawood, Bradd Minnis and Connie Vaughn, American Society for Industrial Security, offered to provide professional security advice to the Task Force.

Appendix C — Task Force Meeting Summaries

9. Paul Barty, Alliance Consulting, suggested a global organization accountable for tracking those that should or need to be tracked.
10. Brian Sherin, DSH Connect, spoke to web-based training for screeners and other airport personnel.
11. Mark Zellers, Aracom, suggested a wireless, broadband solution for incident management.
12. Phil Roberts, Unisys, spoke of making positive I.D. a requirement through iris scanning or fingerprinting.
13. Ronald Martin, Network Alliance, said it was necessary to still use a human resource as a part of the security solution.
14. Mike Cash, Ideaz, spoke in favor of thumbprints as positive identification.
15. James Long, SpectraTek, spoke in favor of video technology for surveillance.
16. David Akers, Eagle Check, Ltd., spoke of a process for security using existing I.D. systems such as Social Security numbers and driver s licenses.
17. Don Treichler, International Brotherhood of Teamsters, Airline Division, spoke of the need for airport workers to have identification and cargo screening.
18. Hal Etterman and Thomas Stoker, Ortega Info Systems, demonstrated a virtual security operations center.
19. Tsahi Gozani, Ancore Corporation, proposed using a scanner to measure the elemental composition of the material content of objects.
20. Scott Lewin & Ken Thorrison, L&M Co., discussed a device to alert security personnel and airport personnel the moment an alarm goes off.
21. Paul B. Barty, Alliance Consulting, discussed his organization s role as security consultants.
22. John J. Deveer, Headland Technologies, spoke in favor of biometric technology.
23. Steve Preminger, South Bay AFLCIO, spoke of the necessity for better wages, training and benefits for airport screeners.

Chair Thompson extended the invitation to anyone else who wished to speak. There being no more presenters, the public hearing was adjourned 3:20 pm.

Appendix C — Task Force Meeting Summaries

Minutes of May 31, 2002
BLUE RIBBON TASK FORCE

For a tape recording of the meeting,
please contact Cindy Kunesh at 408-501-7669

6. **Welcome by Chairman John W. Thompson**

7. **Attendance**

Chairman John W. Thompson, Symantec
Beatriz V. Infante, Aspect Communications
Mike Fox, Sr., M.E. Fox & Company
Thomas Anthony, TSA, Task Force Advisor
Capt. Dan Ashby, Airline Pilots Association/United Airlines
Bill Coleman, BEA Systems
William Crowell, Cylink Corporation
Chief Bill Lansdowne, San Jose Police Department
Krish Panu, @Road
Tino Cuellar, Stanford University
Richard Palmer, Cisco Systems

Other:
Ralph Tonseth
Phil Fok
Frank Jesse
Guy Morgante
Matthew Bostick
Douglas Sabo
Meri Maben
Matthew Bostick
Jim Peterson
Callie Grant
Adam Rak
Charlie Felix
Jim Webb
Sunny Claggett
Darrell Dearborn
Brian Finean
Cindy Kunesh

3. **Minutes Approval**

Chair Thompson called for a motion to approve the Minutes of May 10th. Chief Lansdowne motioned to approve the minutes and Mike Fox, Sr. seconded the motion. Motion carried unanimously.

4. **Subcommittee Reports**

a) Technology Subcommittee

Beatriz V. Infante announced that they were close to having recommendations wrapped up and a summary will be made available at the next meeting. She stated that there were over 40 proposals submitted to the subcommittee. Six proposals were selected and will be presented today. In choosing these six technologies, the subcommittee's focus was primarily on prevention and the integration of information.

b) Report Writing Subcommittee

MIKE FOX, SR. REPORTED THAT THE SUBCOMMITTEE HAS MET SEVERAL TIMES AND AN EXTENSIVE DOCUMENT IS ALREADY IN THE WORKS. SAM ARAKI WILL BE AT THE DOT NEXT WEEK AND WILL BE ABLE TO PROVIDE MORE INPUT FROM THAT MEETING.

5. **Technology Presentations**

a) Bill Cawlfield, Sales Director, Datastrip

Datastrip offers a system based on a high capacity, compact two-dimensional symbology, which offers a cost effective method of capturing and storing secure information such as text, photographs and biometrics, in an area measuring 5/8 x 3 inches that can be printed on substrates such as ID cards, passports and drivers' licenses.

Additionally, a data strip could be encoded on any document that contains the original information from the document; thus, by checking the data strip at the receiving end you could verify that the original document has not been tampered with.

Appendix C — Task Force Meeting Summaries

b) Steve Campano, Marketing Director, Intevac

Laser Illuminated Viewing and Ranging (LIVAR) is a range-gated, laser-illuminated, two-dimensional imaging system that operates in the "eye-safe" wavelength band at 1.5 micron. The eye-safe nature of the laser illumination enables LIVAR to be used in any commercial or military environment with no restrictions on use of the system. It integrates low power infrared/thermal and LIVAR technologies for use in target detection and identification at long ranges.

Electron Bombarded Active Pixel Sensor (EBAPS) is a low light level digital video camera that integrates state-of-the-art night vision performance with a mega pixel digital video camera. The camera incorporates image compression technology to allow low bandwidth transmission and storage of retrieved imagery.

c) Mani Chandy, PhD., Chief Scientist, iSpheres

iSpheres persistently monitors and fuses disaggregated information to identify and respond to security threats in real time. The event-driven software enabling this solution is based on the patent-pending Infospheres Distributed Object System. This distributed system architecture and framework was invented to leverage existing information networks and accommodate large numbers of users and hierarchies.

The iSpheres system continually extracts, parses, and normalizes non-uniform data, looking for matches to specified events or patterns of events. Once identified iSpheres takes action, automatically executing specified responses that can range from alerts/alarms or more complex orchestrations of services such as running an automated contingency analysis and allocate necessary resources.

d) Colin Britton, CTO and Karen Cummings, VP Marketing, Metatomix

Metatomix technology unlocks disparate silos of information and automatically collects the data to discover emerging trends, patterns, and opportunities and/or potential threats. The product is built around a schemaless database called the Hologram Store, which caches data from a variety of sources located in a variety of locations across an enterprise for use in business visibility applications. It can automatically identify common elements between various data, creating new associations and building a more robust, 3-D view of the data a hologram. When a bioterror attack is detected, the system automatically generates an alert and notifies proper officials by phone, fax, email and pagers.

e) Bill Stuntz, President, Broadware (Northrop partner)

Northrop Grumman technology offers network-based digital video solution that allows users to securely view, manage, and store real time live video from anywhere in the world using a standard Internet browser. Unlike traditional digital video records, this system was engineered from the ground up to operate in an IP network environment, employs an open architecture to enable expansion and integration with other systems, and uses COTS technology to increase reliability and reduce costs.

f) Kent Greenough, Vice President, ProActive Implementations Corporation

ProActive provides a 3-D scanning technology that captures large facilities and/or buildings in a "Point Cloud" format. Each point has an x,y,z coordinate; accurate to 6mm. From this enabling technology, ProActive can develop accurate models for use In planning, designing, prototyping and testing a variety of security systems including, perimeter security, access control systems, video surveillance systems, blast mitigation and fire suppression. In addition, the "Point Cloud" imaging can be used in a heads-up display for first responders and can further be developed Into a fully interactive, WAN/LAN training simulator for protection forces, SWAT teams and security forces.

6. **Adjourn to Technology Demonstration Open House for review of Task Force member and subcommittee technologies, including:**

   Cylink
   Identix
   @Road
   Sanctum
   Recognition Systems/Ingersoll Rand
   Symantec
   InVision
   Aspect

Appendix C — Task Force Meeting Summaries

Minutes of June 4, 2002

BLUE RIBBON TASK FORCE

SJC Offices

1732 N. First Street, San Jose, CA

8:30 a.m.

**For a tape recording of the meeting, please contact Cindy Kunesh at 408-501-7669*

1. **WELCOME BY CHAIRMAN JOHN W. THOMPSON**
2. **ATTENDANCE**

Chairman John W. Thompson, Symantec
Sam Araki, Security Technology Ventures
Capt. Dan Ashby, Airline Pilots Association/United Airlines
William Crowell, Cylink Systems
Mike Fox, Sr. M.E. Fox & Co.
Beatriz V. Infante, Aspect Telecommunications
Chief Bill Lansdowne, San Jose Police
Richard Palmer, Cisco Systems
Larry Wansley, American Airlines
Perry Weigle, Sanctum

Other:
Ralph Tonseth
Jim Peterson
Franco Tedeschi
Meri Maben
Jim Webb
Adam Rak
Guy Morgante
Garry Barnett
Gareth Owens
Phil Fok
Leslee Coleman
Callie Grant
Sunny Claggett

3. APPROVAL OF THE MINUTES

Chair Thompson asked the task force members about any additions, deletions or comments to the minutes from the previous meeting held on May 31. There being none, he called for a motion to approve the minutes. The motion passed unanimously.

4. ANNOUNCEMENTS

Chair Thompson reviewed the events that led the committee to its final meeting. They included the work by the Technology Demonstration Committee, including the presentations and exhibits on May 31, and the efforts of the Writing Committee to create a document representative of the group s recommendations. The Chair announced that the process was on schedule for the June 17 presentation of the Task Force findings to Congressman Honda and Mayor Gonzales.

The Chair called upon Sunny Claggett to outline the organization of the June 4 meeting. Sunny explained that there would be an presentation regarding the overview of the report by Larry Gerston, principal writer, followed by a Technology Demonstration Committee report by Beatriz V. Infante, committee chair. After the report, the group would then discuss any issues not covered either by the overview or Technology Demonstration Committee Recommendations.

5. REPORTS

Organization of the Final Report

Larry Gerston provided a sense of the Task Force findings and direction. The final report would emphasize the importance of the application of current technologies to the airport and air travel experiences, focusing upon a validated workforce, a validated facility, and a validated communications structure. Recommendations for these areas would be forthcoming from the Technology Demonstration Committee, with the expected passenger benefits described in each instance. The final report also would emphasize the process through which the task force reached its conclusions, noting the lengths to which the group provided an open, inclusive process.

Technology Demonstration Committee Findings

Beatriz V. Infante took the task force through the research, findings and recommendations of the Technology Demonstration Committee. The committee s areas and recommendations were as follows:

Validated Workforce

1. biometric authentication
2. workforce management

Validated Facility

1. video monitoring
2. driver/vehicle authentication
3. GPS devices to monitor vehicle traffic
4. Access control within the aircraft

Appendix C — Task Force Meeting Summaries

Validated Communications Infrastructure

1. integrated communications structure
2. migration to networked, digital technology

Beatriz showed the task force the ways in which each of the recommendations would provide better protection for passengers, while respecting their civil liberties.

Upon listening to the reports, Sunny facilitated a group discussion on the expect flow of report information as well as any issues not included, but which should be added. At the end of the discussion, the group decided that the report should include mention of the airplane captain s role in the section dealing with validated facility, discussion of the TWIC identification concept in the Unresolved Issues section, and mention of concerns related to cargo in the Unresolved Issues section. Discussion of the task force report and its contents was then declared closed by Chairman Thompson.

APPROVALS AND DEADLINES

Chairman Thompson called upon Larry Gerston to provide a series of deadlines in order to get the report to press on time. Based upon previous consultation with the task force leadership, Larry stated that the final report would be distributed to committee members no later than midnight of June 6th. Members would have until 8:00 p.m. on June 8th to reply with any recommendation changes. Any substantive changes would be cleared with the various committee chairs or representatives. Larry would provide a final report ready for printing no later than 9:00 a.m., June 10. With this schedule, the committee would be confident of a published document in time for the June 17.

ADJOURNMENT

The Chair thanked the committee for its hard work conducted over such a short time. He informed the members that they would be advised of the details regarding the June 17th presentation.

Meeting adjourned at 11:30 a.m.

APPENDIX D

BLUE RIBBON TASK FORCE

Public Hearing
May 10, 2002
1:30 pm

Silicon Valley Conference Center
2161 North First Street
San José, CA

Chair Thompson opened the public hearing by stating that the mission of the Task Force was to develop and submit a report to the Transportation Security Administration in June, outlining recommendations of technologies that can improve security and the security process at our nation's airports. It is also important, he noted, to have the traveler experience improved as well. In light of 9/11, the focus of the Task Force's effort is on three fundamental attributes: cost, convenience and security. The critical element of this mission is to cast the broadest net possible in gathering information and knowledge of the challenges facing aviation security today. Providing the opportunity to hear what the public has to say about these issues will play a key role in ensuring that the efforts are thorough and thoughtful, he said.

Mr. Thompson announced that it is assumed by the Task Force that any information that a presenter submitted here today, is not proprietary information and that it can become a part of the public record. He also said that if someone had an interest in submitting a RFI based upon the submission deadlines that have been set, that RFI submission should be in by midnight, May 10' 2002.

Following is a complete listing of the presenters with a brief note on the subject of their testimony:

1. Steve Kirsch demonstrated new iris scanning technology.
2. Richard L. Rowe, self-employed, presented his ideas on biological sensor technology.
3. William H. Dunlop, Lawrence Livermore National Lab, advised on radioactive sensor technology, which is currently being deployed in Russia.
4. Katie E. Corrigan, American Civil Liberties Union, Legislative Council on Privacy Issues, spoke on privacy, equality and fairness.
5. Rod Dewell, Excalibur Solutions, Inc., a private pilot and biometric technologist, spoke on applying biometric technology in the cockpit.
6. Michelle Kraus, Ping I.D., spoke for digital identity.
7. Helal R. Omeira, Council on American-Islamic Relations, also spoke of privacy, equality and fairness.
8. Jim Cawood, Bradd Minnis and Connie Vaughn, American Society for Industrial Security, offered to provide professional security advice to the Task Force.

Public Hearing Comments – May 10, 2002
Appendix D

9. Paul Barty, Alliance Consulting, suggested a global organization accountable for tracking those that should or need to be tracked.
10. Brian Sherin, DSH Connect, spoke to web-based training for screeners and other airport personnel.
11. Mark Zellers, Aracom, suggested a wireless, broadband solution for incident management.
12. Phil Roberts, Unisys, spoke of making positive I.D. a requirement through iris scanning or fingerprinting.
13. Ronald Martin, Network Alliance, said it was necessary to still use a human resource as a part of the security solution.
14. Mike Cash, Ideaz, spoke in favor of thumbprints as positive identification.
15. James Long, SpectraTek, spoke in favor of video technology for surveillance.
16. David Akers, Eagle Check, Ltd., spoke of a process for security using existing I.D. systems such as Social Security numbers and driver's licenses.
17. Don Treichler, International Brotherhood of Teamsters, Airline Division, spoke of the need for airport workers to have identification and cargo screening.
18. Hal Etterman and Thomas Stoker, Ortega Info Systems, demonstrated a virtual security operations center.
19. Tsahi Gozani, Ancore Corporation, proposed using a scanner to measure the elemental composition of the material content of objects.
20. Scott Lewin & Ken Thorrison, L&M Co., discussed a device to alert security personnel and airport personnel the moment an alarm goes off.
21. Paul B. Barty, Alliance Consulting, discussed his organization's role as security consultants.
22. John J. Deveer, Headland Technologies, spoke in favor of biometric technology.
23. Steve Preminger, South Bay AFLCIO, spoke of the necessity for better wages, training and benefits for airport screeners.

Chair Thompson extended the invitation to anyone else who wished to speak. There being no more presenters, the public hearing was adjourned 3:20 pm.

APPENDIX E

Responses to the Request for Information

Appendix E--Response to the RFI -- Tracking Sheet

Responses to the Request for Information --- Tracking Sheet

Technology Demonstration Committee

Company	POC	Title	RFI	Reviewed	Submission Date
AMCG	Ron Graziosi	Marketing Exec	Yes	Yes	29-Apr
Ancore	Dr. Tsahi Gozani	CEO	Yes	Yes	29-Apr
ArrayComm	Katie Juran	Dir, Communications	Yes	Yes	30-Apr
BaggageDirect	Steve Quackenbush	CEO	No	Yes	18-Apr
Commerce Events	Anand Das	CEO	Yes	Yes	30-Apr
Congruity	Dale Anderson	CEO	Yes	Yes	30-Apr
Convansys	Paul Sumrall	Acct Exec	Yes	Yes	30-Apr
Cylink	Peter Vogt	Dir, Bus Dev	Yes	Yes	30-Apr
Dcard	Jeffery Lui	Pres.	Yes	Yes	18-Apr
GE Interlogix	Mark Duato	Dir, Home Land	Yes	Yes	30-Apr
GeoMetric	Arthur Zwern	CEO	Yes	Yes	19-Apr
ITVR	John Scott	CEO	Yes	Yes	25-Apr
Northrop Grumman	Dena Knuth	IT Acct. Mgr	Yes	Yes	29-Apr
Propel	Steve Kirsch	CEO	No	Yes	29-Apr
SEP Associates	Shaw Pender	Consultant	Yes	Yes	29-Apr
Sony, Government	Robert Wyler	GM	Yes	Yes	29-Apr
Warehouse	Curtis Kent		Yes	Yes	29-Apr
Process Physics	Peter J. Dusza	Pres.	Yes	Yes	10-May
BEA Systems	Heather Dickinson	Public Relations	Yes	Yes	10-May
LochIsle Inc.	Gavin McLintock	Pres.	Yes	Yes	10-May
Modulant	Steve Bastasini	VP of Business Dev	Yes	Yes	10-May
BOLT Systems, Inc.	Sarah Diggs	CEO	Yes	Yes	10-May
DETECTION SUPPORT SERVICES	Michael L. Wantz	Executive Director	Yes	Yes	10-May
Aether Wire & Location, Inc	Bob Fleming		Yes	Yes	10-May
Excalibur Solutions, Inc.	Rod Dewell	Principal Engineer	Yes	Yes	10-May
Lewin & Morrison Enterprises	Scott Lewin		Yes	Yes	10-May
Siemens USA	Ellen Williams	Siemens Airports	Yes	Yes	10-May
Milvac	Rohit Shah		Yes	Yes	10-May
iSpheres Corporation	Robert Naify		Yes	Yes	10-May
Unisys Corporation	Mr. Philip D. Roberts	VP and Managing Principle	Yes	Yes	10-May
Western Disaster Center	Richard H. Davies	Executive Director	Yes	Yes	10-May
ideaz	Mike Cash	General Manager	Yes	Yes	10-May
Network Appliance	Ronald C. Martin	Federal Account Executive	Yes	Yes	10-May
ProActive Implementations Corp	Kent Greenough	Vp	Yes	Yes	10-May
TouchSafe International, Inc.	John M. Cockerham		Yes	Yes	10-May
AVCOM Technologies, Inc.	Brad Bishop	CEO	Yes	Yes	10-May
Intevac, Inc	Steve Campano	Marketing Director	Yes	Yes	10-May
Metatomix, Inc.	Karen Cummings	VP Marketing	Yes	Yes	10-May
Datastrip Inc.	Robert Molina		Yes	Yes	10-May
Voquette, Inc.	Zahoor Kareem	Dr. Business Development	Yes	Yes	10-May
StereoGraphics Corporation	Kevin McCarthy	Dr. Business Development	Yes	Yes	10-May

Bold = May 31, '02 Presenter

APPENDIX F

Technology Demonstration Committee Presenters and Exhibitors—May 31, 2002

F-1

Appendix F – Technology Demonstration Committee Presenters and Exhibitors
May 31, 2002

Schedule

Presentations & Q & A

9:00-9:15 – **Datastrip, Inc.** – Exton, PA
Bill Cawlfield, Sales Director

9:15-9:30 – **Intevac, Inc.** – Santa Clara, CA
Verle Aebi, President, Photonicis Technology Division
Steve Campano, Manager Marketing/Sales

9:30-9:45 - **iSpheres Corporation** – Oakland, CA
Mani Chandy, PhD, Chief Scientist

9:45-10:00 – **Metatomix, Inc.** – Waltham, MA
Karen Cummings, VP Marketing
Colin Britton, Founder and CTO

10:00-10:15 - **Northrop Grumman Corporation,** -- Los Angeles, CA
Bill Stuntz, President, Broadware (Northrop partner)

10:15-10:30 - **ProActive Implementations Corporation,** -- Rancho Cordova, CA
Kent C. Greenough
Kris C. Greenough
Paul Sumrall, Covansys

Exhibitors 5-31-02

Company	POC	Title
Cylink		
Identix	Michael Harvey	Director, Product Marketing
@Road	Rod Fan	CTO
Sanctum		
Recognition / IRCO	Martin Huddart	General Manager
Symantec		
Invision	Tyler Philips	Product Manager
Network Associates		
Aspect Communications	Gareth Owens	Principle Sales Engineer

APPENDIX G

Technology Demonstration Committee Findings

Technology Sub-Committee Recommendations

Silicon Valley Blue Ribbon Task Force on Aviation Security & Technology

Agenda

- Goals
- Scope and Prioritization
- Process Flows / Critical Risk Areas
  - Workforce
  - Passenger
- Technology Blue Print
- Conclusion

CITY OF SAN JOSE
CAPITAL OF SILICON VALLEY

Subcommittee Members

Thomas Anthony	Transportation Security Administration
M. Sam Araki	Security Technology Ventures, LLC
Captain Dan B. Ashby	Air Line Pilots Association International
Matthew Bostick	Office of Congressman Honda
Sandra England	Network Associates
Brian Finan	Symantec Corporation
Michael E. Fox, Sr.	M.E. Fox & Company Incorporated
Don Harris	Southwest Airlines Co.
Beatriz Infante	Aspect Communications
Deborah Magid	IBM
Jim Peterson	San Jose International Airport
Richard W. Palmer, Jr.	Cisco Systems
Krish Panu	@Road
Jim Scott	Recognition Systems
Jim Scullion	Identix.com
Ralph G. Tonseth	San Jose International Airport
Peggy Weigle	Sanctum Inc.
Bill Withycombe	FAA Western Pacific Region

Task Force Goals

- Safety is overriding concern
- Consumer efficiencies
- Workforce efficiencies
- Solutions that leverage existing technology
- Technology vision for the future

Transportation Landscape

Scope

- Multiple Types of Security
  - Physical Security
    - Facilities
    - People
      - Passengers
      - Employees
  - Communications & Data Security
    - Infrastructure
    - Cyber Security

CITY OF SAN JOSE CAPITAL OF SILICON VALLEY

Definitions of Technologies Applicable to Scope

- Physical Security
  - Facilities
    - The ability to effectively screen, identify, and/or authenticate entry, access, clearance and movement of materials and people within a facility.
  - Passenger / Workforce Biometrics
    - Systems that automate in real time the methods of establishing someone's identity from their unique physiological or behavioral characteristics.

Definitions of Technologies Applicable to Scope

- Communications & Cyber Security
  - Communications
    - The seamless exchange of voice and data among databases, networks, applications, and devices that support the synchronization and delivery of real time information.
  - Cyber Security
    - The ability to effectively prevent unauthorized access to sensitive information and protect data among computer networks, applications, and devices.

Prioritizing Principles

- Key Areas of Risk
  - Workforce
  - Physical
  - Communication & Data
- Key Areas of Focus
  - Passenger Convenience
  - Aircraft Access
- Incremental Security
  - High Risk Areas Approached First
- Proactive vs. Reactive Security

CITY OF SAN JOSE CAPITAL OF SILICON VALLEY

Solution Blue Print

Workforce & Passenger Management

Facility Management

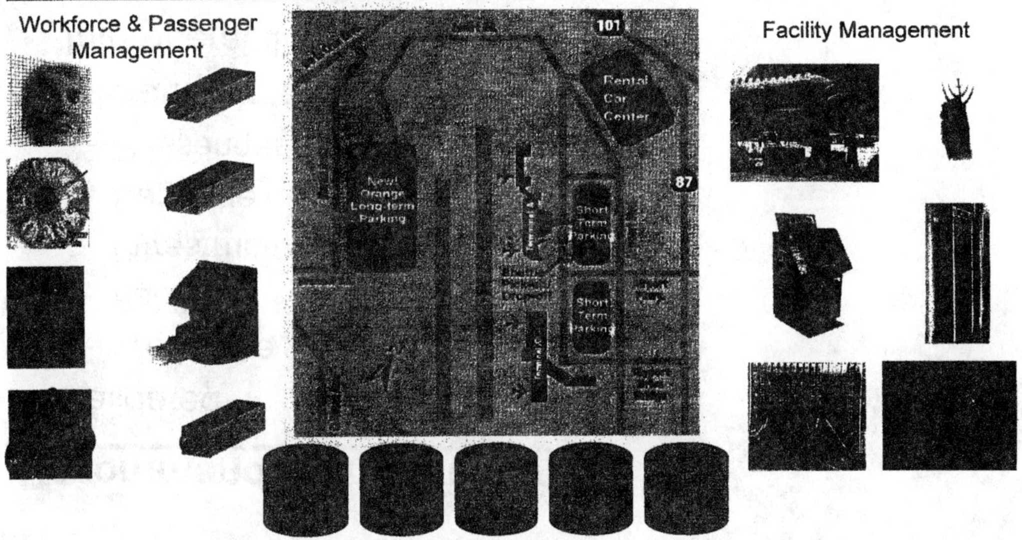

Secure Communications & Information Management

G-11

Recommendation Summary

- Validated
  - Workforce
  - Facility
  - Infrastructure
  - Baggage
  - Passenger
- Migration from Analog to Digital
- Interoperability Across Solutions
- Database Access
- Policy / Civil Liberties

Process Flows

- Employee
- Passenger

Employee Process – Terminal (non-secure)

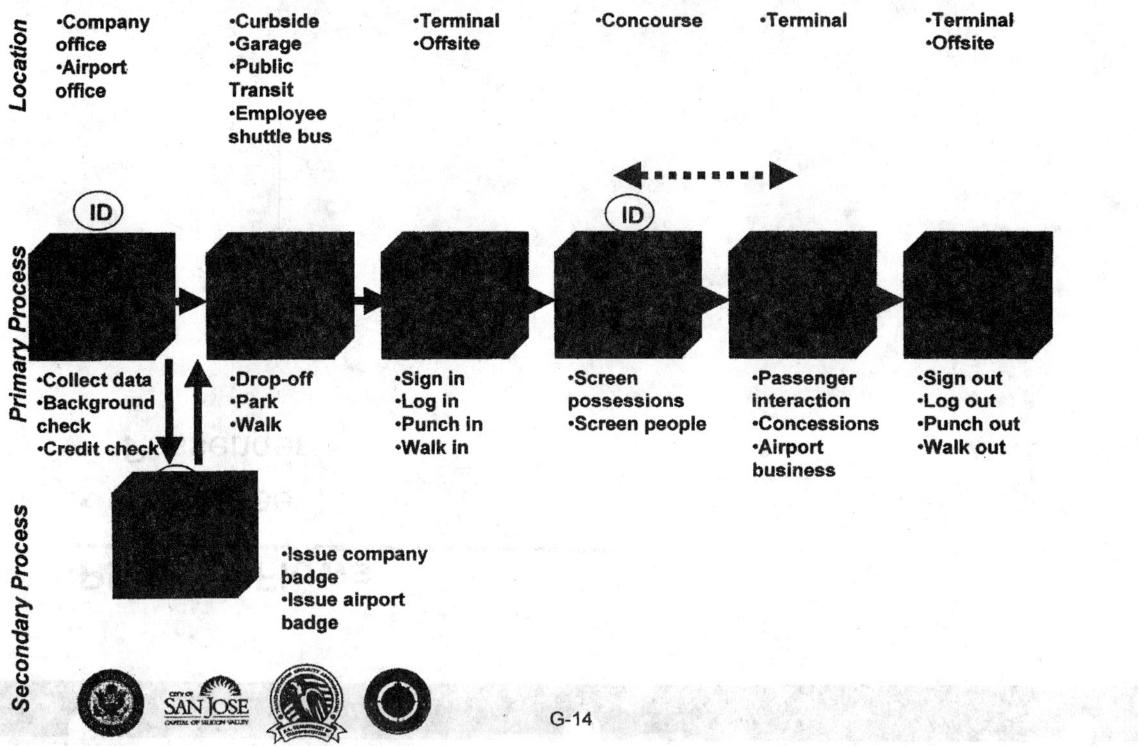

G-14

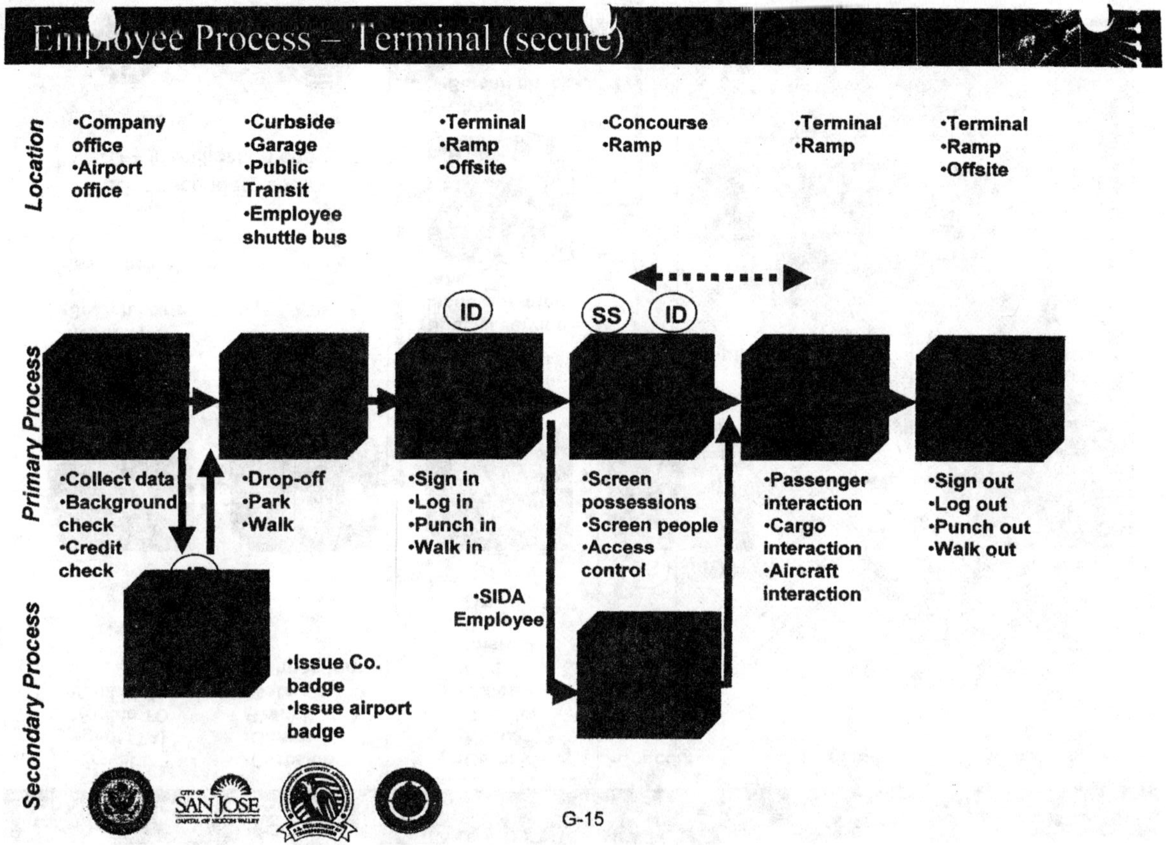
Employee Process – Terminal (secure)
Location
•Company office
•Airport office
•Curbside
•Garage
•Public Transit
•Employee shuttle bus
•Terminal
•Ramp
•Offsite
•Concourse
•Ramp
•Terminal
•Ramp
•Terminal
•Ramp
•Offsite
ID
SS
ID
Primary Process
•Collect data
•Background check
•Credit check
•Drop-off
•Park
•Walk
•Sign in
•Log in
•Punch in
•Walk in
•Screen possessions
•Screen people
•Access control
•Passenger interaction
•Cargo interaction
•Aircraft interaction
•Sign out
•Log out
•Punch out
•Walk out
•SIDA Employee
Secondary Process
•Issue Co. badge
•Issue airport badge
SAN JOSE
G-15

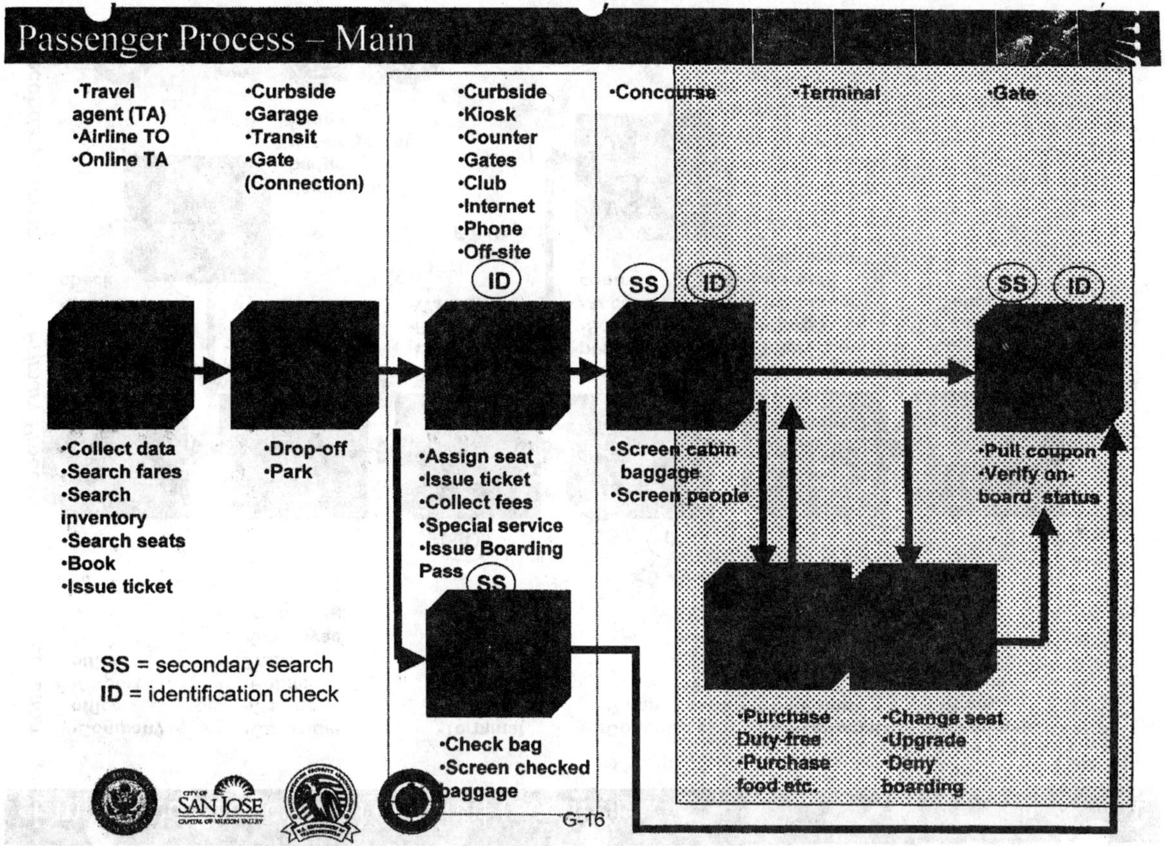
Passenger Process – Main
•Travel agent (TA)
•Airline TO
•Online TA
•Curbside
•Garage
•Transit
•Gate (Connection)
•Curbside
•Kiosk
•Counter
•Gates
•Club
•Internet
•Phone
•Off-site
•Concourse
•Terminal
•Gate
ID
SS
ID
SS
ID
•Collect data
•Search fares
•Search inventory
•Search seats
•Book
•Issue ticket
•Drop-off
•Park
•Assign seat
•Issue ticket
•Collect fees
•Special service
•Issue Boarding Pass
SS
•Screen cabin baggage
•Screen people
•Pull coupon
•Verify on-board status
SS = secondary search
ID = identification check
•Check bag
•Screen checked baggage
•Purchase Duty-free
•Purchase food etc.
•Change seat
•Upgrade
•Deny boarding
G-16

Passenger Process – Main – International

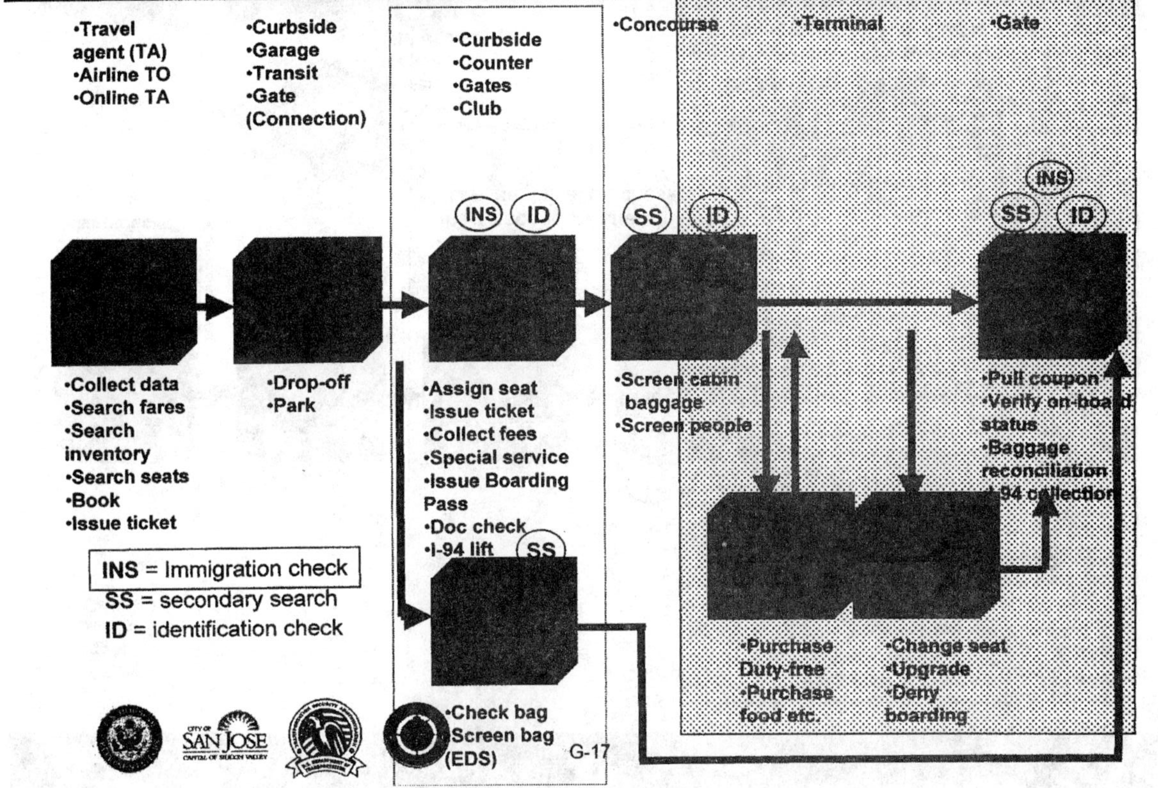

Passenger Process – Off Site

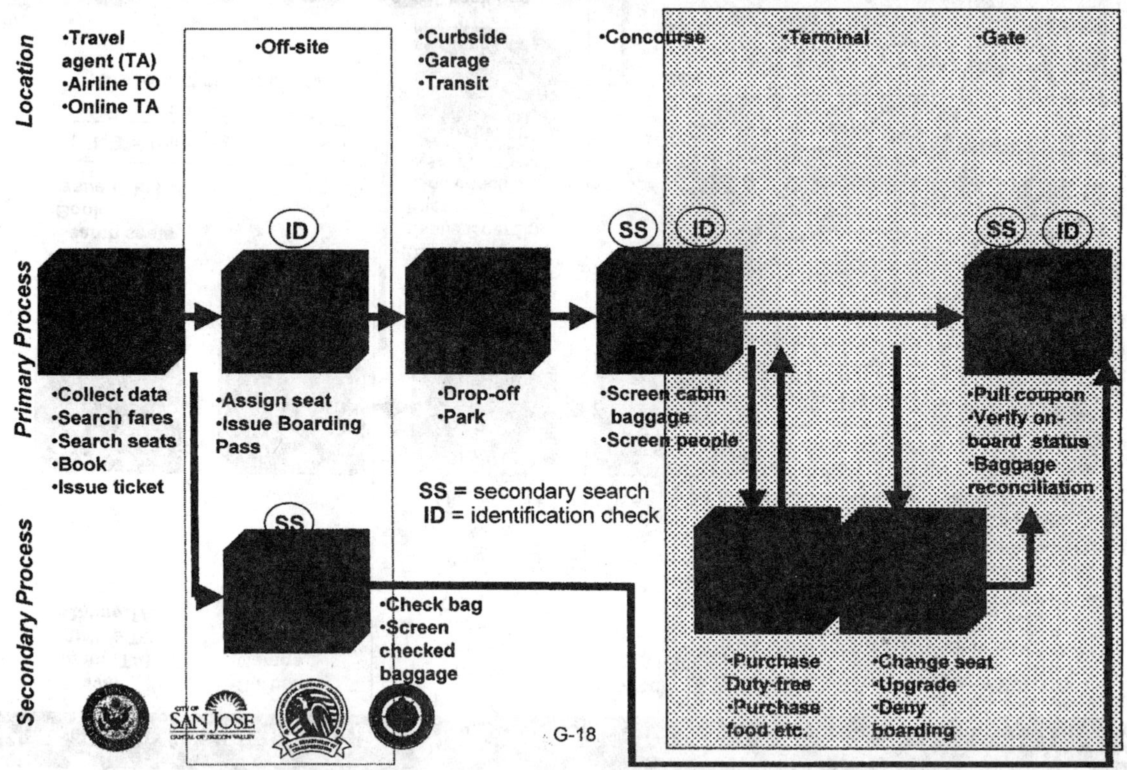

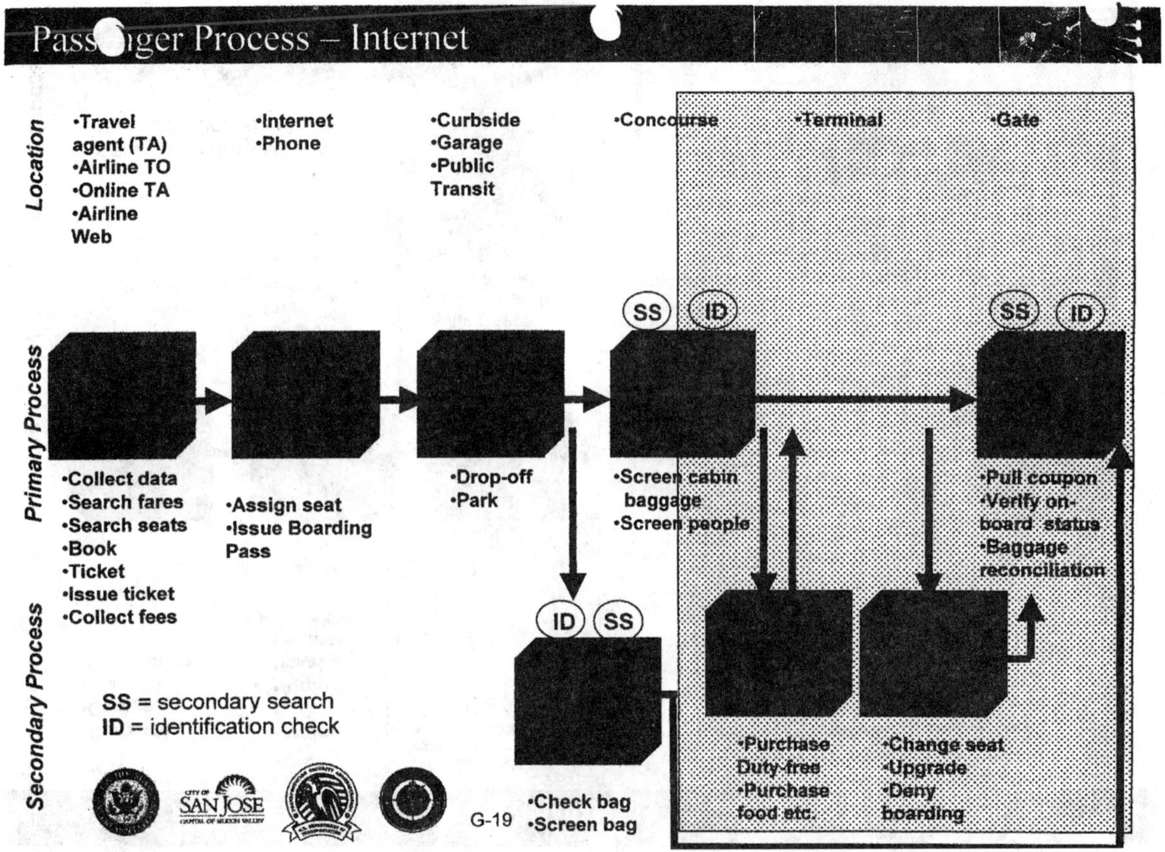
Pass nger Process – Internet
Location
•Travel agent (TA)
•Airline TO
•Online TA
•Airline Web
•Internet
•Phone
•Curbside
•Garage
•Public Transit
•Concourse
•Terminal
•Gate
Primary Process
•Collect data
•Search fares
•Search seats
•Book
•Ticket
•Issue ticket
•Collect fees
•Assign seat
•Issue Boarding Pass
•Drop-off
•Park
SS
ID
•Screen cabin baggage
•Screen people
SS
ID
•Pull coupon
•Verify on-board status
•Baggage reconciliation
Secondary Process
ID
SS
•Check bag
•Screen bag
•Purchase Duty-free
•Purchase food etc.
•Change seat
•Upgrade
•Deny boarding
SS = secondary search
ID = identification check
CITY OF SAN JOSE
CAPITAL OF SILICON VALLEY
G-19

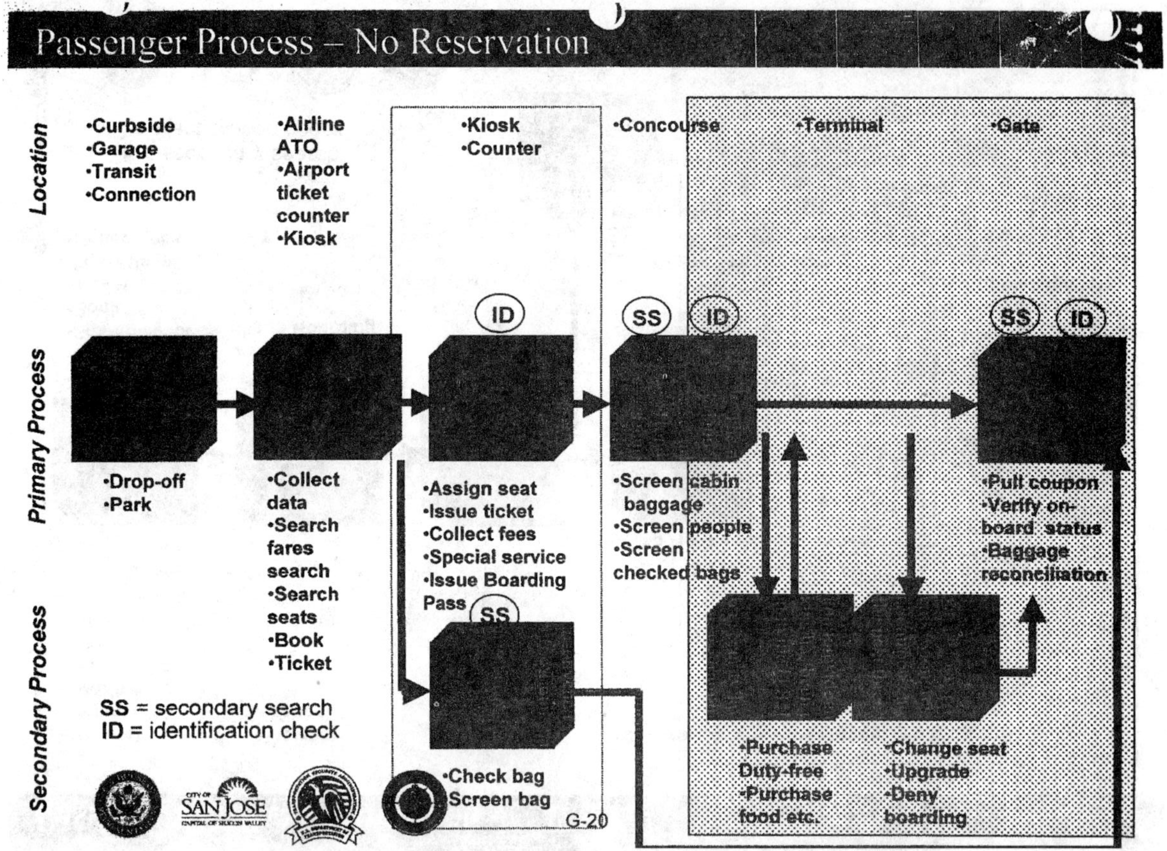
Passenger Process – No Reservation
Location
Primary Process
Secondary Process
•Curbside
•Garage
•Transit
•Connection
•Airline ATO
•Airport ticket counter
•Kiosk
•Kiosk
•Counter
•Concourse
•Terminal
•Gate
ID
SS
ID
SS
ID
•Drop-off
•Park
•Collect data
•Search fares search
•Search seats
•Book
•Ticket
•Assign seat
•Issue ticket
•Collect fees
•Special service
•Issue Boarding Pass
SS
•Screen cabin baggage
•Screen people
•Screen checked bags
•Pull coupon
•Verify on-board status
•Baggage reconciliation
•Purchase Duty-free
•Purchase food etc.
•Change seat
•Upgrade
•Deny boarding
•Check bag
Screen bag
SS = secondary search
ID = identification check
CITY OF SAN JOSE
CAPITAL OF SILICON VALLEY
G-20

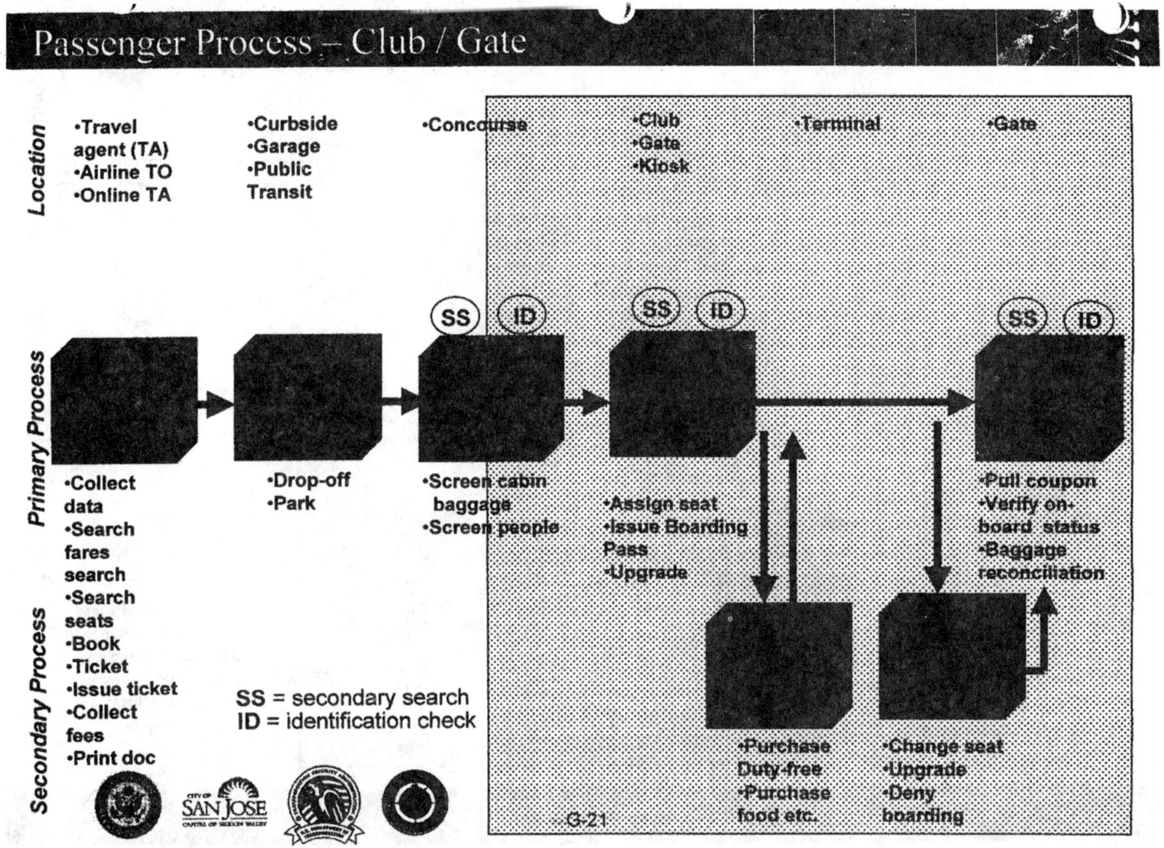
Passenger Process – Club / Gate
Location
•Travel agent (TA) •Airline TO •Online TA
•Curbside •Garage •Public Transit
•Concourse
•Club •Gate •Kiosk
•Terminal
•Gate
Primary Process
SS
ID
•Collect data •Search fares search •Search seats •Book •Ticket •Issue ticket •Collect fees •Print doc
•Drop-off •Park
•Screen cabin baggage •Screen people
•Assign seat •Issue Boarding Pass •Upgrade
•Pull coupon •Verify on-board status •Baggage reconciliation
Secondary Process
•Purchase Duty-free •Purchase food etc.
•Change seat •Upgrade •Deny boarding
SS = secondary search
ID = identification check
SAN JOSE
G-21

Passenger Process – Connection

Location

Arriving international passengers will go to FIS check and re-enter secure area using main process

Primary Process

Secondary Process

SS = secondary search
ID = identification check

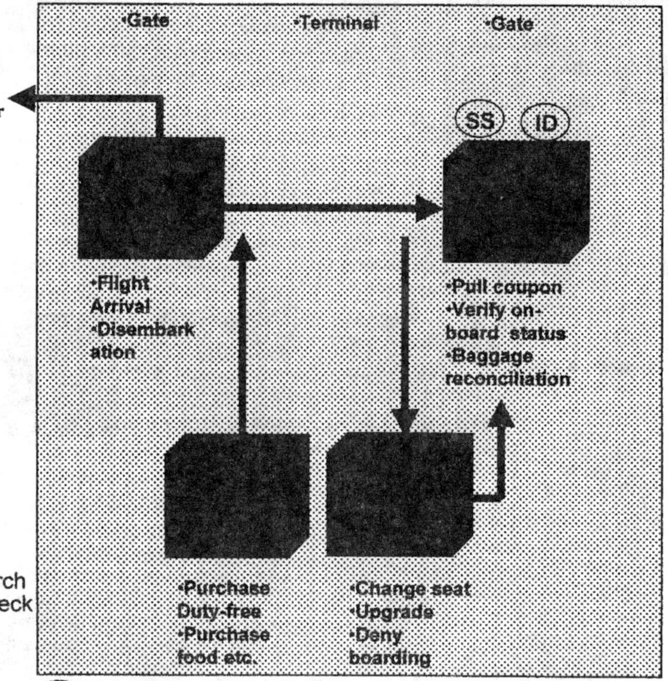

Technology Blue Print

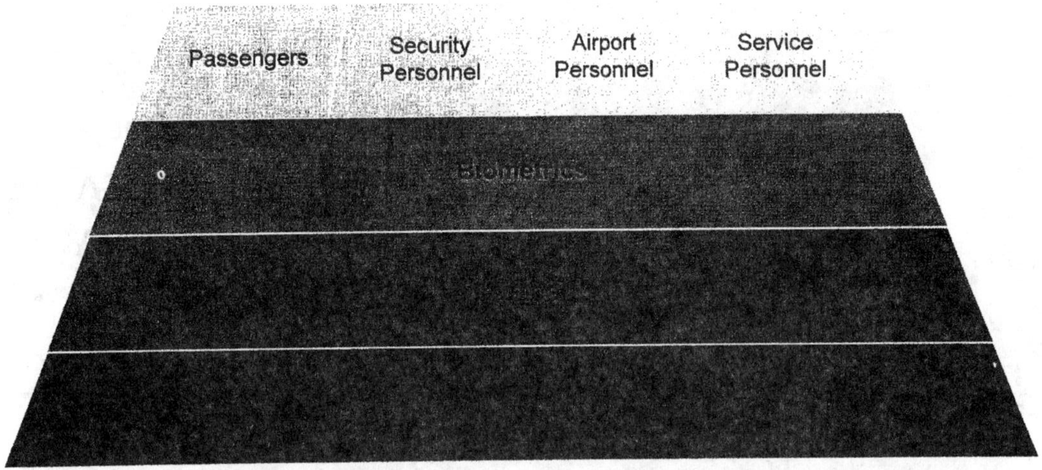

Highest Return on Technology Investment

- Validated Workforce
- Validated Facility
- Validated Infrastructure
- Validated Baggage
- Validated Passenger

Highest Return on Technology Investment #1

- Validated Workforce
  - Workforce Management
    - Annual Background Checks
    - Schedule Tracking
    - Skills Management
    - Alerts & Notifications
  - Digital Identity
    - Intelligent Cards
    - Biometric Capture
    - Alerts & Notifications
  - Video Management
    - Digital Process
    - Access Control
    - Real Time Monitoring and Control
    - Alerts & Notifications

CITY OF SAN JOSE CAPITAL OF SILICON VALLEY

Validated Workforce

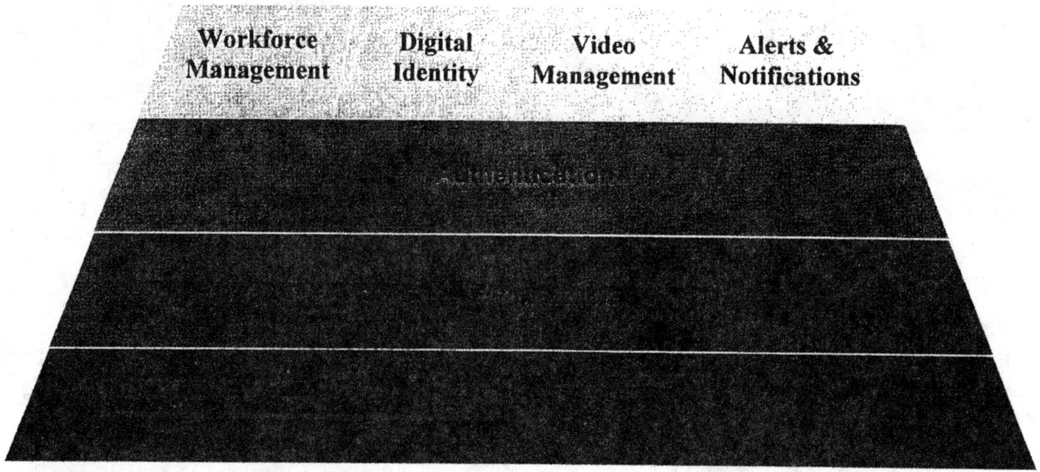

Highest Return on Technology Investment #2

- Validated Facility
  - Workforce Authentication
    - Entry
    - Exit
  - Driver / Vehicle Authentication
    - Driver Biometric Identification
    - Vehicle load inspection
  - Aircraft / Pilot Authentication
    - Pilot Biometric Identification
    - Transport Workers Identification Card (TWIC)

Validated Facility

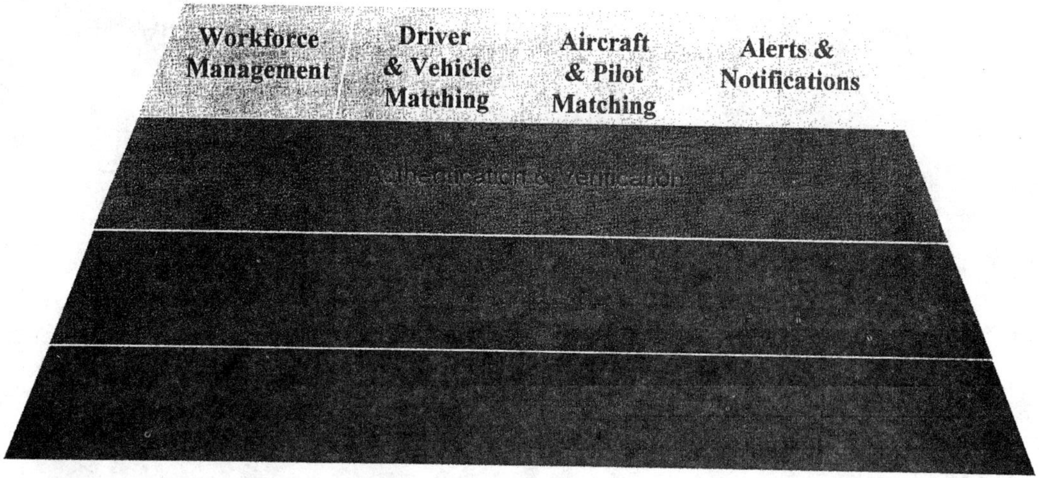

Highest Return on Technology Investment #3

- Validated Infrastructure
  - Common Open Platform and Standards
    - Interoperability
    - Integration point for Physical & Facility solutions
    - Streamlines security & encryption
  - Common Business Rules
  - Migration from Analog to Digital
    - Real time Alerts & Notifications
    - Real time Video
    - Wireless connectivity

CITY OF SAN JOSE CAPITAL OF SILICON VALLEY

Validated Infrastructure

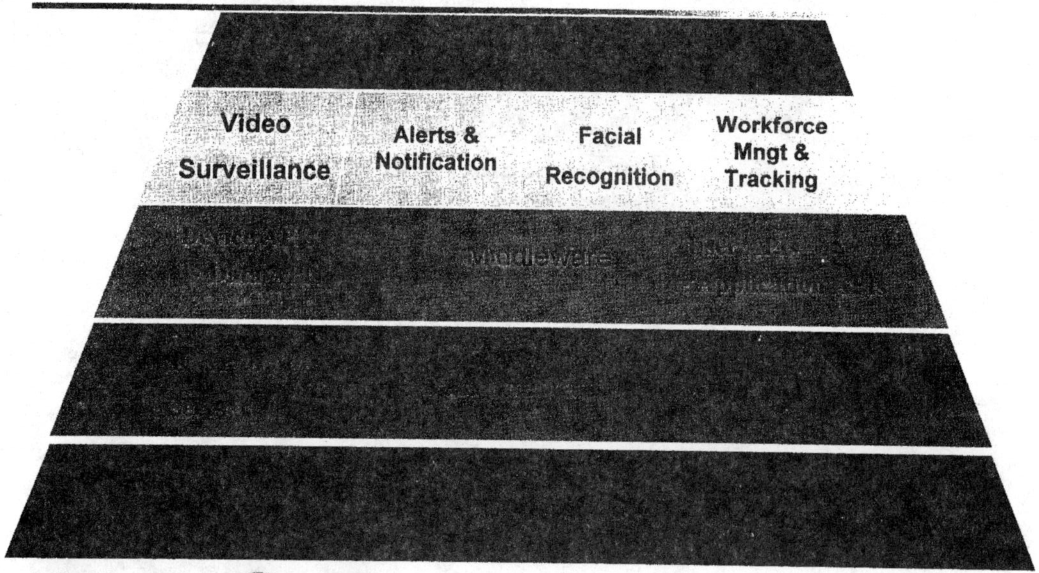

G-30

On-Going Technology Investment

- Validated Baggage
- Validated Passenger

Validated Baggage

- Presently the TSA has issued instructions for baggage inspection. This committee choose to focus on the Workforce, Passengers, Facilities and Communications Infrastructure and let the TSA initiatives move forward without additional confusion.

Validated Passenger

- All technologies used in Validated Workforce can be leveraged
- Key issues are
  - Policy
  - Regulatory
  - Agency

CITY OF SAN JOSE CAPITAL OF SILICON VALLEY

Recommendation Summary

- Validated
  - Workforce
  - Facility
  - Infrastructure
- Migration from Analog to Digital
- Interoperability across solutions
- Policy / Civil Liberties
- Support on-going efforts
  - Baggage
  - Passenger

APPENDIX H

CITY OF
SAN JOSE
CAPITAL OF SILICON VALLEY

MEDIA ADVISORY

For Immediate Release
February 1, 2002

Contact: Ernest Baynard (202) 225-2631 (Honda)
David Vossbrink, (408) 277-3515 (Gonzales)

Rep. Honda, Mayor Gonzales to Establish Blue Ribbon Silicon Valley Aviation Security & Technology Task Force

Event: U.S. Congressman Mike Honda and San Jose Mayor Ron Gonzales will announce the creation of their Silicon Valley Blue Ribbon Task Force to review potential technology solutions to improve aviation and airport security, improve customer service, and provide recommendations to enhance national security.

When: **12:30 p.m.**
Monday, February 4, 2002

Where: **Norman Y. Mineta San Jose International Airport**
Terminal C Media Conference Room
(Next to Mexicana Airlines check-in counter. Validated parking for media is available in Terminal C short-term parking lot)

Who: U.S. Congressman Mike Honda
Mayor Ron Gonzales

Background: The events of September 11 have focused the nation on the need for greater security of airports and the aviation system. Congress and the FAA have recently enacted new regulations for air travel safety, and all U.S. airports are required to implement new security measures by specific dates over the next three years.

The Airport Security Task Force will include technology, security, business, and aviation experts from Silicon Valley. Its goals will be to identify recommendations for innovative and practical solutions to enhance national air travel security and passenger convenience.

Congressman Honda is a member of the House Transportation & Infrastructure Committee, including its Aviation Subcommittee. Honda also serves as a Vice Chair of the Democratic Homeland Security Task Force in the U.S. House of Representatives.

JOSE
SILICON VALLEY

Media Advisory

Office of Mayor Ron Gonzales

mmediate Release:
ruary 1, 2002

Contact:
David Vossbrink, (408) 277-3515
Communications Director

Mayor Gonzales and Congressman Honda to Establish Silicon Valley Aviation Security Technology Task Force

Event: San Jose Mayor and Congressman Mike Honda will announce plans to establish a Silicon Valley Blue Ribbon Task Force to review potential technology solutions to improve aviation and airport security, improve customer service, and provide recommendations to enhance national security.

When: **12:30 p.m.**
Monday, February 4, 2002

Where: **Norman Y. Mineta San Jose International Airport**
Terminal C Media Conference Room
(Next to Mexicana Airlines check-in counter. Validated parking for media is available in Terminal C short-term parking lot)

Who: Mayor Ron Gonzales
Congressman Mike Honda

Background: The events of September 11 have focused the nation on the need for greater security of airports and the aviation system. Congress and the FAA have recently enacted new regulations for air travel safety, and all U.S. airports are required to implement new security measures by specific dates over the next three years.

The Airport Security Task Force will include technology, security, business, and aviation experts from Silicon Valley. Its goals will be to identify recommendations for innovative and practical solutions to enhance national air travel security and passenger convenience.

Congressman Honda is a member of the House Transportation Committee and Aviation Subcommittee.

801 N. First St., Suite 500, San Jose, CA 95110 *tel* (408) 277-5800 *fax* (408) 277-3755 *web* www.sjmayor.org

NEWS RELEASE

CITY OF SAN JOSE
CAPITAL OF SILICON VALLEY

For Immediate Release
February 4, 2002

Contact: Ernest Baynard, (202) 225-2631 (Honda)
David Vossbrink, (408) 277-3515 (Gonzales)

Rep. Honda, Mayor Gonzales Establish Silicon Valley Aviation Security & Technology Task Force

San Jose, CA U.S. Congressman Mike Honda (D-San Jose) and San Jose Mayor Ron Gonzales announced today the formation of a Blue Ribbon Task Force comprised of up to 20 technology, security, business, and aviation experts from Silicon Valley to identify and evaluate technology-driven solutions to improve the security and efficiency of national and local aviation. Once the task force is named, it will have 100 days to develop recommendations regarding existing and emerging technologies that can upgrade systems for passenger identification, baggage screening, airfield and cockpit security, explosive detection and other security concerns. The panel s recommendations will be submitted by Honda and Gonzales to the San Jose City Council and the new head of the Transportation Security Administration, Undersecretary of Transportation, John Magaw.

Since September 11th, I have been privileged to work with Mayor Gonzales, business leaders, and representatives from all levels of government to begin the development of a comprehensive, technology-driven security system to keep our airways safe, said Honda. The Blue Ribbon Task Force will build upon this collaborative effort by drawing upon the rich mosaic of energy and innovation within Silicon Valley to dramatically improve security, efficiency and technology at our nation s airports.

San Jose is an ideal location to explore technology approaches for making our nation s airports and travel safer without adding delays for travelers and freight, said Mayor Gonzales. Our city s commitment to innovation and service has already been successful in making our airport a model for fast and effective response to the new transportation challenges after September 11.

According to Honda, the task force will be ready to begin its work by early March. We expect we will begin reviewing its findings in June, said Honda. This is on a very fast track so that we can share our results with Transportation Secretary Norm Mineta, Undersecretary Magaw, the FAA and other airports and move toward new solutions as quickly as possible.

Gonzales noted the long tradition of successful partnerships in Silicon Valley. We have demonstrated that we can achieve results by working with both the public and private sectors and between local and federal governments, said the mayor. This is another opportunity to take advantage of our practical approach to problem solving for national benefit.

More .

Honda and Gonzales Announce Airport Security & Technology Task Force **2-2-2**

The events of September 11 focused the nation on the need for greater security throughout the aviation infrastructure. Congress and the FAA have enacted broad new regulations for air travel safety, and all U.S. airports are required to implement an array of new security measures by specific dates over the next three years. To ensure uniform and nationwide implementation of these new measures, the responsibility for airport security operations has been placed under the jurisdiction of the new Transportation Security Administration in the U.S. Department of Transportation, headed by Undersecretary of Transportation John Magaw.

The new federal aviation security law includes a key provision authored by Honda that will launch a nationwide pilot program in twenty or more U.S. airports to test new and emerging security technologies. The measure, which is also part of freestanding legislation introduced by Honda and Rep. Jim Matheson (D-UT) in October 2001, could dramatically improve airport security by promoting the development and use of cutting edge technologies, such as biometric authentication, Global Position System applications, enhanced communication systems and database integration protocols. The pilot program is a product of numerous meetings and demonstrations that Honda convened with Mayor Gonzales, top Silicon Valley executives, the FAA, FBI, Bay Area Airport officials and other Members of Congress.

Silicon Valley is in a unique position to design and implement cutting-edge solutions to meet our new aviation security needs in the shortest possible timeframe, added Honda. One of the top priorities for the Task Force will be to help ensure that Mineta San Jose International Airport is selected as a pilot project under the new federal aviation law and so that it can be become a national model for security, efficiency and innovation.

Congressman Honda is a member of the House Transportation Committee s Aviation Subcommittee and also serves as a Vice Chair of the Democratic Homeland Security Task Force n the U.S. House of Representatives.

-30-

NEWS RELEASE

CITY OF SAN JOSE
CAPITAL OF SILICON VALLEY

For Immediate Release
February 27, 2002

Contact: Ernest Baynard, (202) 225-2631 (Honda)
David Vossbrink, (408) 277-3515 (Gonzales)
Cris Paden, (408) 517-8547 (Symantec)

Airport Security Technology Task Force Taps Symantec CEO John W. Thompson as Working Chair

San Jose, CA U.S. Congressman Mike Honda (D-San Jose) and San Jose Mayor Ron Gonzales announced today that John W. Thompson, CEO of Symantec Corporation, will serve as the Chair for the Silicon Valley Blue Ribbon Task Force on Aviation Security and Technology.

The task force will be comprised of up to 20 technology, security, business, and aviation experts from Silicon Valley to identify and evaluate technology-driven solutions to improve the security and efficiency of national and local aviation.

John W. Thompson s expertise and ability make him an ideal choice to head this Task Force. He rightly understands that true aviation security means protecting not only airplanes and people, but networks and data as well, said Honda. **Throughout our history, the public sector and private enterprise have worked together to face our nation s greatest challenges the Blue Ribbon Task Force hopes to build upon this important legacy.**

The diverse and powerful resources of our area make Silicon Valley uniquely prepared to harness technology to improve security and efficiency at our nation s airports, said Thompson. **It is an honor to be selected by Congressman Honda and Mayor Gonzales to work in partnership with our business leaders and elected officials to develop innovative solutions to enhance national aviation security.**

Gonzales noted that Thompson s successful experience leading high technology enterprises and in the field of Internet security will provide a valuable perspective and focus for the task force. **We expect this group to identify where the tools and technology can be developed to help us both improve security and improve air travel convenience at our own airport and at others,** he said. **With John Thompson s outstanding leadership and technical knowledge, I am confident that the task force will provide good results.**

The balance of the task force will be named in the coming weeks. Once the task force is established, it will have 100 days to develop recommendations regarding existing and emerging technologies that can upgrade systems for passenger identification, baggage screening, airfield and cockpit security, explosive detection and other security concerns.

More .

Honda and Gonzales Name Chair of Airport Security Technology Task Force **2-2-2**

The panel s recommendations will be submitted by Honda and Gonzales to the San Jose City Council and the new head of the Transportation Security Administration, Undersecretary of Transportation, John Magaw.

The events of September 11 focused the nation on the need for greater security for aviation infrastructure and systems. Congress and the FAA have enacted broad new regulations for air travel safety, and all U.S. airports are required to implement an array of new security measures by specific dates over the next three years.

To ensure uniform and nationwide implementation of these new measures, the responsibility for airport security operations has been placed under the jurisdiction of the new Transportation Security Administration in the U.S. Department of Transportation.

The new federal aviation security law includes a key provision authored by Honda that will launch a nationwide pilot program in twenty or more U.S. airports to test new and emerging security technologies. The measure, which is also part of freestanding legislation introduced by Honda and Rep. Jim Matheson (D-UT) in October 2001, could dramatically improve airport security by promoting the development and use of cutting edge technologies, such as biometric authentication, global positioning system applications, enhanced communication systems and database integration protocols. The pilot program is a product of numerous meetings and demonstrations that Honda convened with Mayor Gonzales, top Silicon Valley executives, the FAA, FBI, Bay Area Airport officials and other Members of Congress.

Congressman Honda is a member of the House Transportation Committee s Aviation Subcommittee and also serves as a Vice Chair of the Democratic Homeland Security Task Force in the U.S. House of Representatives.

John W. Thompson is chairman of the board of directors and chief executive officer of Symantec Corporation. Since joining Symantec in April 1999, Thompson led the transformation of the company from a consumer software publisher to the global leader in Internet security solutions for individuals and enterprises.

Thompson joined Symantec after a 28-year career with the IBM Corporation where he held senior executive positions in sales and software development. Prior to joining Symantec, he was general manager of IBM Americas with responsibility for sales and support of IBM's technology products and services,

Symantec is a world leader in Internet security technology. The Silicon Valley company provides a broad range of content and network security software and appliance solutions to individuals, enterprises and service providers. Headquartered in Cupertino, Calif., Symantec has worldwide operations in 38 countries. For more information, please visit www.symantec.com.

-30-

NEWS RELEASE

CITY OF SAN JOSE
CAPITAL OF SILICON VALLEY

For Immediate Release
March 13, 2002

Contact: Ernest Baynard, (202) 225-2631 (Honda)
David Vossbrink, (408) 277-3515 (Gonzales)
Cris Paden, (408) 517-8547 (Symantec)

Honda, Gonzales Name Members of Airport Security Technology Task Force

San Jose, CA U.S. Congressman Mike Honda (D-San Jose) and San Jose Mayor Ron Gonzales announced today the members of the Silicon Valley Blue Ribbon Task Force on Aviation Security and Technology.

The task force includes 19 technology, security, business, and aviation experts from Silicon Valley ***[roster attached]***. Their goal is to identify and evaluate technology-driven solutions that will improve the security and efficiency of national and local aviation. John W. Thompson, CEO of Symantec Corporation, will serve as the chair of the blue ribbon group.

These business, technology, and aviation leaders represent the unique capabilities of Silicon Valley, said Congressman Honda. **I am confident they will come back quickly with innovative and practical recommendations that will help make our nation s aviation system safer and easier to use.**

Our goal is to achieve greater safety and shorter lines through innovation, said Mayor Gonzales. **The expertise and commitment of San Jose and Silicon Valley will again lead the way for creative and effective technology solutions to help protect the nation s airports and air travelers.**

I m looking forward to working with an excellent group that has been assembled by Congressman Honda and Mayor Gonzales to serve our region and our nation, said John W. Thompson. **This is a wonderful opportunity for businesspeople to work in partnership with local and federal government toward a common goal that benefits the public and our economy.**

The task force has 100 days to develop recommendations regarding existing and emerging technologies that can upgrade systems for passenger identification, baggage screening, airfield and cockpit security, explosive detection and other security concerns.

More .

Honda and Gonzales Name Members of Airport Security Technology Task Force **2-2-2**

The panel will hold at least one public hearing to provide an opportunity for public participation and to listen to additional suggestions that could be used for improving aviation security. The panel is scheduled to complete its work in June so that Honda and Gonzales can submit its findings and recommendations to the San Jose City Council and Undersecretary of Transportation John Magaw, the new head of the federal Transportation Security Administration.

Members of the task force were recommended to Honda and Gonzales by Silicon Valley business associations including the Information Technology Association of America, Semiconductor Industry Association, Business Software Alliance, Silicon Valley Manufacturing Group, and San Jose Silicon Valley Chamber of Commerce.

The events of September 11 focused the nation on the need for greater security for aviation infrastructure and systems. Congress and the FAA have enacted broad new regulations for air travel safety, and all U.S. airports are required to implement an array of new security measures by specific dates over the next three years.

To ensure uniform and nationwide implementation of these new measures, the responsibility for airport security operations has been placed under the jurisdiction of the new Transportation Security Administration in the U.S. Department of Transportation.

The new federal aviation security law includes a key provision authored by Honda that will launch a nationwide pilot program in twenty or more U.S. airports to test new and emerging security technologies. The measure could dramatically improve airport security by promoting the development and use of cutting-edge technologies, such as biometric authentication, global positioning system applications, enhanced communication systems and database integration protocols.

The airport pilot program is a product of numerous meetings and demonstrations that Honda convened with Mayor Gonzales, top Silicon Valley executives, the FAA, FBI, Bay Area Airport officials and other Members of Congress. The FAA will select the 20 pilot sites by this summer.

Congressman Honda is a member of the House Transportation Committee s Aviation Subcommittee and also serves as a Vice Chair of the Democratic Homeland Security Task Force in the U.S. House of Representatives.

-30-

[Roster on next page]

Honda and Gonzales Name Members of Airport Security Technology Task Force

Silicon Valley Blue Ribbon Task Force on Aviation Security and Technology

Member	Source of Nomination	Affiliation	Title
Mike Honda	Honorary Chair	U.S. House of Representatives	Member of Congress
Ron Gonzales	Honorary Chair	City of San Jose	Mayor
John W. Thompson	Mike Honda Ron Gonzales	Symantec	CEO, Chairman
Sam Araki	Mike Honda	Security Technology Ventures	Chairman, Former CEO, Lockeed
Dan Ashby	Mike Honda	United Airlines/ALPA	Pilot/Chair of California Airline Pilots Assoc.
Bill Crowell	AEA	Cylink Corp.	CEO, President
Tino Cuellar	Mike Honda	Stanford University	Professor, School of Law
Sandra England	BSA	Network Associates	Exec. VP, Business Development & Research
Mike Fox, Sr.	SJCoC	M.E. Fox Distributing	President
Dan Harris	Ron Gonzales	Southwest Airlines	Director, Systems Projects for Ground Ops.
Beatriz V. Infante	AEA	Aspect Communications	CEO
Bill Lansdowne	Ron Gonzales	City of San Jose	Chief of Police
Dr. Sergio Magistri	AEA	InVision Technologies	CEO, President
Bob McCashin	AEA	Identix Incorporated	CEO, Chairman
Ko Nishimura	SIA	Solectron	CEO, Chairman
Richard Palmer Jr.	Ron Gonzales	Cisco Systems	VP, VSEC Business Unit
Krish Panu	ITAA	@ Road	CEO
Larry Wansley	Ron Gonzales	American Airlines	Managing Director for Corp. Security
Tom Weidemeyer	John W. Thompson	UPS	COO-UPS, Pres.UPS Airlines
Peggy Weigle	ITAA	Sanctum	CEO

ITAA: Information Technology Association of America
BSA: Business Software Alliance
AEA: American Electronics Association
SIA: Semiconductor Industry Association
SJCoC: San Jose Silicon Valley Chamber of Commerce

NEWS RELEASE

CITY OF SAN JOSE
CAPITAL OF SILICON VALLEY

For Immediate Release
April 29, 2002

Contacts: Ernest Baynard, (202) 225-2631 (Honda)
David Vossbrink, (408) 277-3515 (Gonzales)
Lakshmi Bakshi, (408) 325-2623 (Infante)

Aviation Security Task Force Extends Deadline For Submission of Products, Ideas to Improve National Aviation Security, Efficiency

Blue Ribbon Task Force extends deadline to May 10

SAN JOSE, CA —The Silicon Valley Blue Ribbon Task Force on Aviation Security and Technology today announced that it is extending the deadline for companies to submit proposals for consideration as the Task Force begins to identify current and emerging technologies for improving security at U.S. airports. Applicants will now have until May 10, 2002 to make submissions to the Task Force.

Companies can submit technology proposals by visiting the Task Force s website at www.sjcblueribbontaskforce.org. The website has been developed to handle responses from technology companies in Silicon Valley and across the nation and help the Task Force meet its deadline to complete its work by June.

The Task Force also will be holding a public meeting on Friday, May 10, 2002, at 1:30 p.m., to provide the opportunity to residents and technology companies to raise concerns and share ideas regarding efforts to use technology to enhance safety and efficiency at the nation s airports. The hearing will be held at the Silicon Valley Conference Center in San Jose.

We extended the deadline due to the high volume of responses we have already received and to allow for more time for additional submissions as a result of the upcoming public meeting, said Ernest Baynard, spokesman for Congressman Mike Honda.

Congressman Honda and San Jose Mayor Ron Gonzales established the Task Force in March to work with Mineta San Jose International Airport and Silicon Valley business and technology leaders in making recommendations than can enhance national aviation security through technology. Symantec CEO John W. Thompson chairs the Task Force.

Thompson recently appointed Beatriz Infante, CEO of San Jose-based Aspect Communications, as chair of the Task Force s Technology Subcommittee. The subcommittee will manage the submission process and work with the Mineta San Jose Airport to forward submissions best suited for further consideration by the Task Force as a whole.

-30-

MEDIA ADVISORY

CITY OF SAN JOSE
CAPITAL OF SILICON VALLEY

For Immediate Release
May 6, 2002

Contacts: Ernest Baynard, (202) 225-2631 (Honda)
David Vossbrink, (408) 277-3515 (Gonzales)

Aviation Security Task Force to Hold Public Hearing On Technology Solutions and Concerns

Event: The Silicon Valley Blue Ribbon Task Force on Aviation Security and Technology will hold a public hearing this Friday to provide the opportunity to the public to raise concerns and share ideas regarding efforts to use technology in order to enhance safety and efficiency at the nation s airports.

When: **Friday, May 10, 2002**
1:30 p.m.

Where: **Silicon Valley Conference Center, 1st Floor**
2161 North First Street (at Brokaw Road), San Jose
Light Rail Stop: Karina

Who: Members of the Silicon Valley Blue Ribbon Task Force on Aviation Security and Technology.

Background: The Task Force was formed in March by Congressman Mike Honda and San Jose Mayor Ron Gonzales to work with Mineta San Jose International Airport and business and technology leaders in making recommendations than can enhance national and local aviation security and efficiency through technology. The Task Force is chaired by Symantec CEO John W. Thompson.

The primary goal of the Blue Ribbon Task Force is to provide support and guidance to the U.S. Department of Transportation by reviewing a wide array of emerging aviation security technology proposals. It will recommend the most promising options to US Secretary of Transportation Norm Mineta, Undersecretary for the Transportation Security Administration John Magaw, the San Jose City Council and other officials. The Task Force is scheduled to complete its work by June 2002.

-30-

U.S. HOUSE OF REPRESENTATIVES

CITY OF SAN JOSE
CAPITAL OF SILICON VALLEY

Public Hearing

Blue Ribbon Task Force on Aviation Security and Technology

Do you have suggestions or concerns regarding how technology can enhance security and improve efficiency of the nation's airports and aviation system?

The Silicon Valley Blue Ribbon Task Force on Aviation Security and Technology is holding a public hearing for this purpose on:

Friday, May 10, 2002, 1:30 p.m.
The Silicon Valley Conference Center, 1st Floor
2161 North First Street (at Brokaw Road), San Jose
Light Rail Stop: Karina

The Task Force is comprised of Silicon Valley technology, security, and aviation leaders, and was established in March by U.S. Congressman Mike Honda and San Jose Mayor Ron Gonzales. The goal of the Task Force is to identify and evaluate technology-driven solutions to improve the security and efficiency of national and local aviation.

For more information, visit www.sjcblueribbontaskforce.org

CITY OF SAN JOSE
CAPITAL OF SILICON VALLEY

Request for Submissions

Blue Ribbon Task Force on Aviation Security and Technology

Does your business have innovative technology products or applications that could be used to enhance security and efficiency at our nation's airports?

You are invited to submit information for consideration by the Silicon Valley Blue Ribbon Task Force on Aviation Security and Technology. The submissions must be received by 5:00 p.m., May 10, 2002. Selected finalists will be invited to make formal technology presentations to the Task Force on May 31, 2002.

The Task Force is comprised of Silicon Valley technology, security, and aviation leaders, and was established in March by U.S. Congressman Mike Honda and San Jose Mayor Ron Gonzales. The goal of the Task Force is to identify and evaluate technology-driven solutions to improve the security and efficiency of national and local aviation.

All submissions must be submitted online. For more information or to make a submission, please visit www.sjcblueribbontaskforce.org